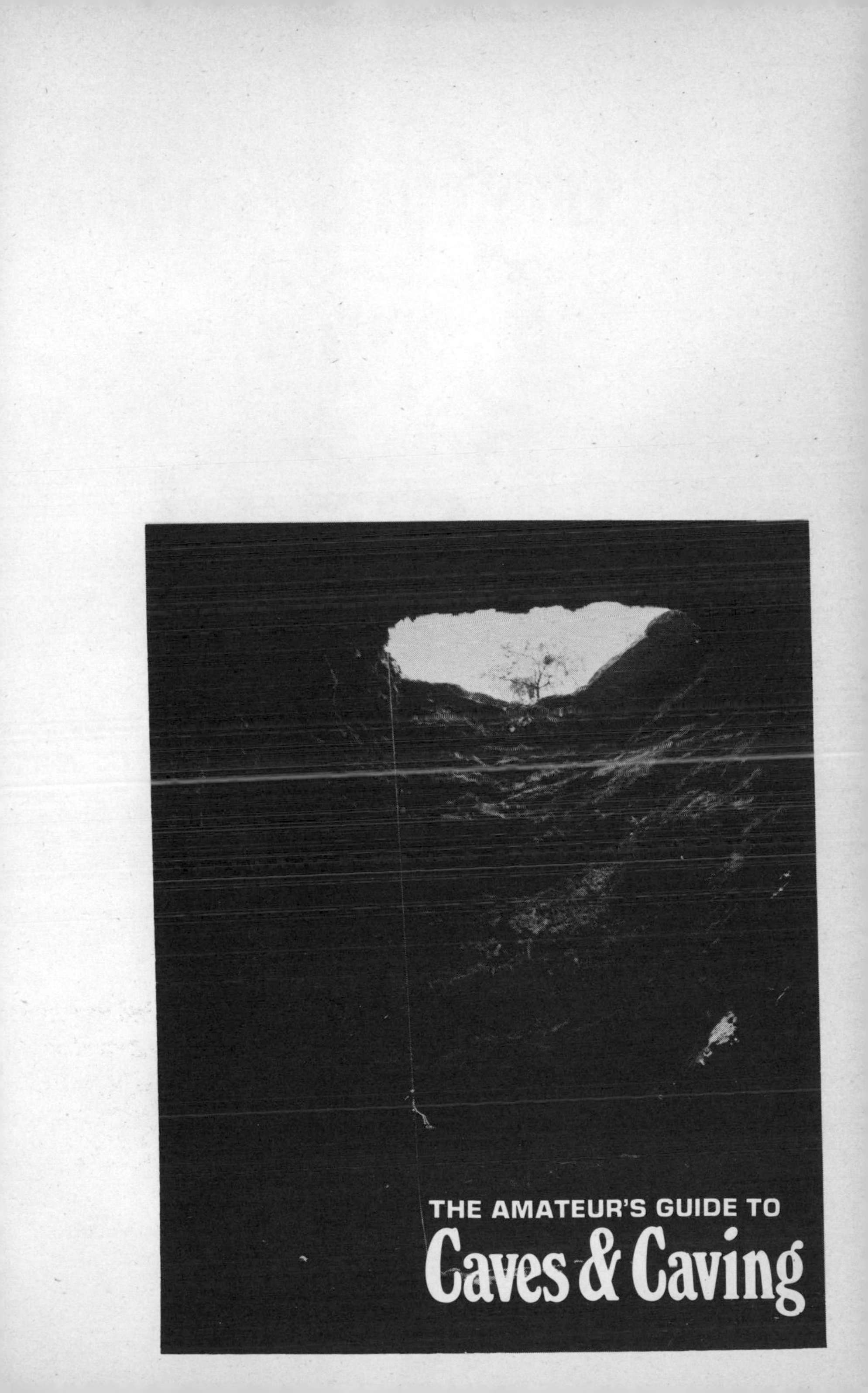
THE AMATEUR'S GUIDE TO
Caves & Caving

The Amateur's Guide to Caves & Caving

DAVID R. McCLURG

FRONTISPIECE: Caver rappelling into 136-foot entrance pit, Devil's Sinkhole, Texas.

(Photo by Charlie and Jo Larson. This and the other photos by them are Copyright 1972, C. and J. Larson, Vancouver, Washington.)

Drawings by LaRhee Parker

skill-building ways to finding and exploring the underground wilderness

ENDORSED BY
THE NATIONAL SPELEOLOGICAL SOCIETY

Stackpole Books

THE AMATEUR'S GUIDE TO CAVES AND CAVING

Published by
STACKPOLE BOOKS
Cameron and Kelker Streets
Harrisburg, Pa. 17105

Printed in the U.S.A.

Library of Congress Cataloging in Publication Data

McClurg, David R.
The amateur's guide to caves and caving.

"Endorsed by the National Speleological Society."
Bibliography: p.
1. Caves. I. Title.

GB602.M32 796.5'25 72-14152
ISBN 0-8117-0094-1

Contents

Preface

I met Dave McClurg in 1960 when he and has family came to settle in California, where I then lived. He joined with us in the activities of National Speleological Society members that included some complex and fascinating explorations of some of California's most interesting caves. In the years since, I have been associated with Dave in other speleological pursuits. I know him as a man fascinated by the cave environment, glad to share his knowledge and feelings about this unique world with others, and as an individual deeply concerned about the many destructive forces that confront the fragile subterranean world.

The National Speleological Society has members and chapters across the whole country and has grown up alongside the development of America's caves. This knowledge has included not only information about the geology and biology of these caves, but also the techniques of cave exploration that ensure safety and comfort, and a strong concern for the protection of these caves from vandalism, pollution or environmental alteration of any form. Dave McClurg has played an important role in these scientific, technical and philosophical developments and is particularly able to convey them to cave explorers-to-be. The Society's endorsement of this book reflects the fact that his guidance to beginners, while very much his own, illustrates the ideals of the Society in regard to the choice of good equipment, the use of good techniques, the practice of high safety standards, and the implementation of good principles of cave conservation.

Dave McClurg has told me that, in writing this book, he regretted that there is no third-person singular in the English language to replace "he" and "his" (there is in Turkish, in the forms *o* and *onun*). Both men and women are involved in the activity of cave exploration and in the study of caves, so women reading this book are not to find anything "sexist" in the pronouns "he" and "his"—otherwise, wait for the Turkish edition. . . .

Are caves dangerous? You will get as many answers as the number of cavers you ask. My answer is yes, they are dangerous for the inexperienced, the careless and the rash. A large portion of this book is Dave McClurg's answer to this same question, in the form of advice on what to do about it. Beyond that, this book is more than a general introduction to caving; it is an experienced caver's recollection of all the lessons he learned in exploring beneath the surface of the earth; what equipment he found essential or useful; how he organized cave trips; what he did and saw; and how he visited this lightless world without damage to himself—or to the cave. As in any other activity requiring knowledge and skills, you won't really learn them entirely from this book, but a book can tell you how another person saw the problems and their solutions, and that is a large part of knowledge. To gain the skills, as Dave McClurg emphasizes, the best way is to travel with experienced cavers.

But there is a dilemma. Imagine a world of near total silence, absolutely without light, of constant temperature and usually without much motion of air or water. That apparent isolation is the cave world to which many living organisms have adapted and by that adaptation have come to depend upon such darkness and quietude. Add to this the exquisite, inanimate, fragile stalactites, stalagmites—even deposits of mud—that have taken centuries to form. We explorers with our lights and noise and fumes and heavy feet are alien intruders. Happily, we can enjoy our visit in little time and then leave the cave world to its own slow, dark tempo—*if we are careful.* The dilemma of cave exploration is to enjoy our visits to what seems "another world," and to leave that world unaltered, unspoiled, for all time.

Dave McClurg explains the need for cave conservation and points out the practices that will help ensure the protection of caves and everything, living and inanimate, that they contain, but only experience with caves, and with those who look upon them as veritable *living museums,* can make these attitudes come alive for the newcomer to cave exploration. I hope that this book leads the reader to those first experiences in caves that will convey the desirability, and problems, of preserving these unique undergound wildernesses.

I have dwelt upon the delicacy of caves and their contents because this is the one aspect of caves least known or understood by most people. But this does not mean that most caves cannot be visited; it only argues for the special care that is now becoming second nature to the present day caver, as shown by the many cautions expressed by Dave McClurg. So, if anyone reading this book is tantalized by the prospect of cave exploration, I urge them to learn all they can from this book, to seek knowledgeable and experienced companions, and to discover for themselves the mysteries, the beauty, the pleasures, and the perils of caves.

Rane L. Curl, *President*
National Speleological Society
Ann Arbor, February 1973

1

What is Caving?

Caves—what does that word bring to mind? Mystery, adventure, discovery, beauty, conservation, danger. To many who are avid cavers and speleologists, caves are all of these things and many more, too.

Does the word "caves" conjure up these kinds of visions in your mind? Then perhaps that spark of interest that often ignites into enthusiasm has been kindled already and questions like these will have come to mind:

"Is caving dangerous? What's it like going underground for the first time? Why do you go caving? Would I like it? How can I help preserve the underground beauty?"

From the story of our first visit to a wild cave the answers to some of these questions may become apparent. But before that, it might be best to make it clear just what this book is and who it was written for.

This is a book about *caving,* not a book about caves or cavers! In other words, it is a basic introductory manual for the would be caver, not a description of famous caves nor the exploits of well-known cavers. It will tell how to get started, what equipment to take, and generally how to learn the techniques needed. It will help a tyro find an organized group (such as a chapter of the National Speleological Society), so he can cave safely and get the proper training. It will also explain the basic facts about cave conservation to let him enjoy caves more and help preserve them for others.

However, it must be said from the outset that caving is not an armchair sport. In caving, the action is underground. Just reading about caving is like going out for football and then spending the whole season on the bench. To be a caver, the aspirant must learn his basic skills in the actual cave environment in the company of an experienced group.

Caving is also dangerous. One new at caving could memorize a book like this word for word, and still be unsafe underground. He would have gained some appreciation for the dangers involved, but without actual practice in a cave with experienced cavers he would still stand a good chance of injury or death. However, with the proper equipment and training, caving is no more hazardous than other high risk activities like water skiing, scuba diving, rock climbing, or sky diving.

Besides being a manual for beginners, this book can serve as an excellent refresher for experienced cavers. If any readers count themselves in this category, they need merely check the chapter headings for areas of particular interest to see if anyone has come up with some new ideas. Check too, the chapter on conservation and safety to be sure everyone is right up to date on the current thinking of organized caving.

If there are any conservationists who are wondering whether a book like this won't perhaps "overadvertise" caving, let this be said now: Any book by itself, including this one, isn't going to increase the total number of cavers very much. In fact, its main appeal will mostly be to those who have already caved a little, anyway. Furthermore, those who read it will come away more aware and better informed about caving and cave conservation. So in the balance, it makes sense to have a book available to those new to caving that will give them the proper safety and conservation message in the first place.

A FIRST CAVE TRIP

Now—about that first underground experience that answered so many questions for us about caving: It happened like this:

We were driving in the country one Saturday afternoon in an area noted for its many caves. There were seven of us: a young couple from the local chapter (grotto) of the National Speleological Society who were going to be our guides, our three pre-teen and teen-aged children, my wife, and myself. As we were driving along, one of the children saw an entrance to a cave over on a hillside.

"At least it looks like an entrance," he said. "Can we stop and see?"

We had visited several commercial caves, but this day would mark our first encounter with the underground wilderness. Our first task was to locate the owner who, it turned out, lived at the farmhouse we had just passed.

"Sure, you can take a look at the cave," he said. "Just go slow and take it easy, there's nothing to be afraid of. You can put your car over by the barn. But leave the gates as they be!"

"Thank you," answered our caving guides. "We've been in before on an NSS grotto trip. We'll check with you as soon as we come out."

With our newfound enthusiasm barely in check, we walked over to the entrance located at the base of a small hill. It was about eighteen inches wide and four feet high. It looked dark and foreboding. Our guide said he would go first and his wife would bring up the rear, putting all of the beginners in the middle for safety.

One by one we squeezed through the entrance, stooping down slightly to avoid bumping our heads. We were struck immediately with how valuable those hardhats were that our caving friends had loaned us. Just inside, it was cool, quiet, and very dark. It was also rather frightening, although none of us would admit it. Soon we became conscious of two distinctive smells. The first was that earthy smell of dirt that comes from being literally *inside* the earth. The second was an acrid smell that was new to us then, but which we soon came to associate so closely with caves. This was the telltale smell of acetylene from the carbide lamps mounted on the front of our hard hats.

There was a pause for a moment to let our eyes get used to the darkness, but soon we were ready to push on and see what lay beyond in the twisting, downward-sloping passage. Movement was slow to avoid stumbling because the passage was covered with irregularly shaped rocks that looked like pieces from a crazy jig-saw puzzle. They were muddy and quite slippery. Besides, many were loose and moved when stepped on, so we had to test each one before putting our weight down firmly. It was a lot like picking a way across a stream on the small rocks in the stream bed, trying hard to avoid falling in the water. In this passage there was no stream (as there often is in caves), but because of the limited light and our inexperience, there was always the danger of twisting an ankle or taking a bad spill.

We moved quite slowly, the narrow fingers of lamp light piercing the darkness as they swept the three foot span between the damp walls, first down to the floor, then back up to the six foot ceiling as we inched along in the darkness. So far, we novices were all having a wonderful time. So much so, that no one noticed when the daylight streaming through the entrance, a last link with the outside world, could no longer be seen behind.

Soon we came to a room with two branching passages, like a Y. On closer examination, it was apparent that there were really three branching passages here, affording a choice of going ahead to the left or right in the Y, or trying the third passage coming in at a very steep angle from almost behind. At the time we didn't particularly notice this latter passage.

"Let's always take the left hand passage when there's a choice," suggested our oldest daughter. "That way we can always take the right hand passage when we come back out and we won't get lost."

Our guides agreed that this seemed a good plan, so to the left it was. At this point we had actually been in the cave about a half an hour and, as is common with the underground experience, we had completely lost track of time. In fact, we were, as cavers often are, suspended in time and space in a dark, seemingly friendly but always potentially hostile environment.

If it is true, as the psychologists say, that to really relax one must leave his normal world behind, then perhaps this is one of the major attractions of caving. It is like entering a world totally different from anything our experience has led us to expect. Dark, quiet, cool, dangerous, unchanging. These are the qualities found in abundance in the underground wilderness.

Meanwhile one of the children had moved ahead with the guide and shouted back that they had found a large room.

"Hey you guys, this is really a big room. It must be at least three hundred feet across and seventy feet high."

As the rest of us filed into the room and began to shine our lights into the dark void, we realized that he wasn't exaggerating much. It certainly was a very large room. In the darkness of a cave, particularly after moving down a relatively narrow passage, large rooms often seem far larger than they really are. This is one of the things that helped add color and romance to the early write-ups about caves in the late nineteenth century.

Entering this new part of the cave seemed to break the spell a little and it was decided that we should think about turning back. But first, a good look at the far wall of the room, which seemed to be the end of the cave. With the help of our two caving friends, the children checked out a couple of small muddy passages in this far wall. We waited at the mouth of these passages for safety's sake, always a good practice when checking out new leads. However, it turned out that these apparently didn't go anywhere and so we were even more convinced that this would be an excellent time to turn back and bring our first adventure to an end.

What none of us could know then was that within a few months, a group of local NSS cavers would return to this room and routinely check out these small passages again. This time, however, they would discover a connection missed by hundreds of cavers before them—a connection that eventually took them into another four and a half miles of cave, with deep pits, a good sized lake, and a maze of passages, many of which were beautifully decorated with glistening speleothems: soda straws, canopies, columns, and even some of those rare gravity-defying helectites! With this new discovery, a nearly forgotten beginners' cave suddenly became the object of great interest and study by cavers and speleologists who had thought its secrets as well as most of its original beauty were long gone.

Simple though this cave may have been before the new discovery, we still managed to have one brush with danger that caused a minor moment of panic on the way out. The guides had asked if we thought we could find our way from the big room to the entrance.

"Of course," said the children in concert, "nothing to it. We just take the right hand passage every time."

We were doing just fine until we eventually came back to the junction room where the three branching passages and the entrance passage all came together. We hadn't paid much attention to its appearance on the way in, and in particular, we hadn't done what seasoned cavers do as a matter of routine: turn around to see what the passage looks like when coming from the other direction. This simple expedient is one of the easiest ways to keep from getting lost in a cave.

When we came back into the junction room, we unhesitatingly

took the passage to the right and suddenly found ourselves moving deeper and deeper into an area we didn't recognize at all. Fortunately, we had only gone a few hundred feet before one of the children said:

"I don't think we've been here before. This doesn't look familiar."

With that, we all stopped to cool our nerves a bit. This time, a little more wisely, perhaps, we sent the older children back down the passage to see if they could get to a point where they recognized something. Almost immediately they found themselves in the junction room again and said:

"Hey, this is it. Come on, let's find our way out."

It was true. This was the way out. When we got back to the junction room we realized that this time the passage to the left was the one we should have taken. Within only a few minutes, we made a sharp turn and found that the world was still out there: we saw the welcome sight of sunlight streaming in through the entrance. Our guides were smiling as we all tumbled happily out into the sunlight. They too had been lost on their first time in, since, the local NSS grotto invariably uses that junction room as a training exercise for all their beginners.

Speaking from our present vantage point as experienced and sophisticated cavers, that trip seems like a long time ago. But still, it's one that none of us will ever forget. I wonder if you have already had that kind of experience, and perhaps that's the reason you're reading this book. Or perhaps you have been intrigued by visits to commercial caves and wondered what lay beyond the lights of the commercial tour—what lay in those dark passages in the caves beyond.

2

Conservation and Safety

As many people both inside and outside of caving have come to realize, conservation is a lot more than just a convenient list of do's and don'ts. It is a state of mind—a belief that some natural features of this earth are worth saving because they help man know more about himself and broaden his increasingly narrow view of nature.

In common with many natural features, caves are especially fragile. Underground beauty that took millions of years to develop can be destroyed in seconds by the thoughtless or accidental acts of cavers and non-cavers alike.

Please keep this in mind. Every beginning caver must be ready to join with others in observing the simple conservation practices needed to save caves. If he is not willing to make this personal commitment, he should stop now and seek another sport or activity. There is no doubt that the few caves of the world can't accept any more cavers unless they are fully committed to saving and preserving what's left of the underground wilderness.

American and Canadian cavers are fortunate to have the benefit of a unified organization devoted to caving and cave conservation, the National Speleological Society (NSS). Unlike the smaller groups and factions of cavers in most other areas, this Society can speak with a single voice on the subject of caves and cave preservation.

In the NSS, there is a simple motto that sums up its feelings on conservation:

"Take nothing but pictures;
Leave nothing but footprints;
Kill nothing but time."

CONSERVATION RULES

Inside a cave, the rules of conservation are simple. Don't *leave* anything and don't *take* anything. This means a caver never leaves trash, spent carbide, or batteries. Instead, he takes them home and throws them in his own trash. In particular, don't bury carbide or batteries in a cave or on a farmer's land. Too many animals have died from this thoughtless act.

Don't remove any speleothems. It doesn't matter if they are already broken off. If a caver wants a record of a particular formation, he takes a picture of it. (Slides are always popular at caving meetings.) Even broken speleothems can tell a story that only a geologist can interpret. The same goes for leaves, tree branches, or artifacts of any kind. Leave them *"in situ."* A scientist can only use them if they are undisturbed.

The entrance area just outside a cave is a common dumping ground for broken speleothems that somehow didn't seem as pretty out it daylight as they did in the cave. Remember, if one breaks them off, he's destroying them forever. To a speleologist, a vandalized cave is like a museum in which someone has broken all the display cases and scattered the contents all over the floor, stealing whatever appeals to him, and ruining even the walls with names and obscenities.

When a caving group comes to an especially beautiful area in a cave, the members should be careful, particularly on the way out when they are tired. If any damage would occur, experienced groups invariably choose an alternate route.

Don't collect animal or plant life in caves except under the specific direction of a scientist. He will provide the proper containers and outline the correct methods of handling so the sample will have value to him. If one finds what he thinks may be of interest to a speleologist, tell him about it first. He may already have a specimen, hence taking another may upset the delicate life chain in the cave unnecessarily.

Several years ago, it was common to smoke arrows on the walls of caves with carbide lamps indicating the way out in junction rooms or confusing places. No responsible caver marks a cave wall any more, with arrows, names, or whatever. In complex mazes, bring along Scotchlite arrows or other reminders that can be removed upon departure. (Surveyors sometimes leave tiny numbered tabs or paint small indicators to mark permanent stations. This seems an acceptable compromise, if done carefully.)

Cave conservation is a must, for if those who visit the caves the most don't preserve them, then no one else will. Those delicate soda straws and elf-like gypsum flowers, those blind fish, flying mammals called bats, and pure white salamanders will become as extinct as the terradactyl.

This is one case where conservation is really up to the individual. Cavers are the ones who use the caves. And if they misuse them they will destroy within their own generation natural beauty that took tens of millions of years to develop!

CAVE SAFETY

Caving is dangerous, make no mistake about that. But it is also true that caving is only as dangerous as a caver makes it. A properly trained and equipped caver is usually safer underground than a careless (or tired) driver on the freeway.

Caving requires teamwork. This is another way of saying that the first rule of safe caving is *never cave alone*. What is a relatively minor accident when with an experienced group, such as a failing lamp, a twisted ankle, or a broken rope, is often fatal to the solo caver. There is no doubt that an easy way to get killed is to enter a cave alone with only a single flashlight for illumination, a ball of string to find the way out, and a knotted piece of clothesline for any pits encountered. Some cavers go so far as to say that this isn't just stupid, it's a clear case of suicide.

Danger is always present in caves, just as it is in mountain climbing, sailing, scuba diving, and private flying. To some, this aspect of caving is one of its main attractions. But no one courts danger deliberately. Rather, by good technique, training, and equipment, danger can be overcome. It is this fact that provides deep satisfaction to many cavers.

Although several cavers have been killed in underground accidents in recent years (including several NSS members), the total is not nearly as high as in mountaineering and other high risk sports. However, this should lull no one into sloppy caving. The danger is always there and must not be underestimated.

Four Cavers Minimum

If a caver shouldn't cave alone, then what is the minimum number for a safe cave trip? The answer is four. Many once considered three experienced cavers as the minimum, but four is far safer.

Here's the reason why. In case of an accident, four cavers are needed so that one can stay with the injured caver, while the other two go for help. (Remember, when help is needed, the same never-cave-alone rule still applies!) With beginners in the party, there should be at least one experienced caver to every three beginners. However, never have less than two experienced cavers on any trip with beginners, so one can lead and one can bring up the rear.

On any type of cave trip, some kind of leader is necessary. With beginners, this is mandatory. However, experienced cavers often don't actually designate a leader for every trip. Instead, they seem to understand instinctively the unspoken caving rule which states: the most experienced or strongest caver is the leader and he will assume command if a leader is needed because of some unexpected danger or accident.

Leaving Word

When going off on a cave trip, the individual should always tell someone about his plans, including the name of the cave, what parts he plans to visit, and when he expects to come out. Allow several hours leeway on the exit time. It never hurts to come out early, but somehow when a caving party meets its rescuers determinedly slugging their way in on a mission of pure mercy, their would-be saviors never see the humor of the situation.

The best person to leave word with is the cave owner or caretaker at the time his permission is sought. If there is no one home, leave a note. If it is known ahead of time that no one will be there or no one lives nearby, one should leave word back at home with family or a fellow caver. No matter who it is, be explicit about plans, so that rescuers, if any are ever needed, can find the group with a minimum of time and effort.

Hard Hats and Lights

No experienced caver ever goes underground without a hard hat equipped with a chin strap, and at least three sources of light. In fact, a caver is never seen without his hard hat, which is why a group of them emerging from a crawlway often resembles a bunch of cycloptic earthworms wriggling happily in the mud.

A caver's three sources of light commonly work out to be a main carbide or electric lamp on the helmet, a flashlight with cord attaching it to coveralls, and candles with matches in a waterproof container. But some cavers have been known to carry three carbide lamps, or an electric headlamp and two flashlights. The magic number is at least three, and the exact combination is less important than the quantity. With whatever type of main light, always carry spares for twice the expected stay—extra carbide, water, and spare parts, or extra batteries and bulbs.

Exceeding Capabilities

Knowing what one can do and can't do are also essential to safe caving. A beginning caver should never be afraid to say that he wants a safety line or doesn't feel he can make a certain climb. If he's tired or doesn't feel well that day, he shouldn't go in at all. Or if already in the cave, he should tell the others. They will respect him for it, because a reckless or inconsiderate caver doesn't just endanger himself, he is a real menace to the whole group. If he should injure himself by a careless move or a foolhardy stunt, it's the others who will end up caring for him and who may have to carry him out—no small task in cave passages and pits. Caving is a team activity, and this includes knowing and never exceeding one's limitations.

Mines—Stay Out

A word about mines for the beginning caver. Never explore mines alone or even with a so-called experienced group. Mines are about as near to death traps as one would ever want to come. Whereas caves were formed millions of years ago and have been subject to countless earthquakes and similar forces, mines are only tens or at most hundreds of years old. Falling timbers, bad air, passage collapse—these are a few of the ways to get killed.

Cave Air

With only one or two exceptions in the entire world, caves never contain bad air. In fact, many caves breath on a regular cycle as the outside barometric pressure changes. But bad air can be encountered for short periods of time in rare cases where there are large quantities of decaying vegetation or pollution from nearby gasoline or chemical storage tanks.

If a novice notices that he and the others in the group seem to be all getting unaccountably short of breath in a cave, the best thing to do is to get out fast. Lighting a candle can provide a crude test for bad air, but one shouldn't waste time for tests unless he has just noticed the symptoms of bad air. Contrary to popular opinion, a carbide lamp is not a good test. It will continue to burn after the danger point to humans is reached. If the test even partially confirms one's suspicions, leave the cave. The effects of bad air are swift and deadly.

Underground Fires

Never light a fire underground. On longer trips where hot food is recommended, bring in a small backpacker's stove. Bat guano is flammable and explosive. One instance is recorded of a fatal explosion and fire that burned for two years in a guano-filled cave. Besides, it's not very smart to light a fire that could use up the air supply and replace it with deadly carbon monixde. Fires also pollute cave walls and formations.

Lost?

Getting lost in caves is a much rarer occurrence than one might think, but nearly every caver (or at least every honest one), will admit to having been "momentarily confused" once or twice in complex cave systems.

If one ever does get lost, or if he loses his light for a moment, he shouldn't panic. The worst thing to do is to go running headlong down a passage. Instead, one should sit still for a moment and collect his senses. Light a candle and turn the main light down or off completely. Give a loud yell at regular intervals to see if the others will answer. Keep it up even if not hearing a reply. Sometimes because of intertwining passages, sound travel seems to be a one-way proposition.

Believe it or not, a lost caver will be found sooner than he expects, and he'll be none the worse for wear—if he stays put and doesn't lose his head.

Cave Flooding

Occasionally, the lower passages or even an entire cave will flood during the rainy season. This is a real danger in some parts of the country and the risk cannot be taken lightly. The rule is, don't go into caves with underground streams if the skies are dark or if it has rained a lot in the past few hours or days. Even though it may seem to be clearing up, the flood water may already be coursing its way underground and in several hours may reach the cave or the passages one desires to enter.

When underground, be alert for any changes in air movement, rising stream or lake levels, increased noise from a stream or foam at the base of a waterfall, or increased debris and mud in the water. If any of these warning signs are observed, start immediately for the entrance or get to high ground there or in an upper passage.

Play it safe when flooding is suspected. Don't risk adding to the fatalities already recorded by those who have ignored the obvious warning signs provided by nature.

Entering a New Cave

When with a group that is the first to enter a recently discovered cave or section of an older cave, remember that the thrill of discovery has its dark side, too. Unstable ceiling rocks can be set loose by a vibration as small as that from a voice, let alone from a jar caused by a caver trying to wedge through some breakdown blocks. Go slowly and quietly. Watch out, too, for dangerous floors of thin calcite under which the dirt or gravel has been washed away. It may be a false floor actually many feet above a real floor level below.

Upon encountering an unsafe looking cave or room, tip-toe out quietly. Then leave a warning note at the entrance and tell others about the unstable condition.

Remember also, the responsibility of conservation. Choose routes carefully to cause as little damage as possible in new caves.

Old Ladders and Ropes

One word of caution which bears repeating several times: Never use ladders, ropes, bolts, or any other mechanical devices found in a cave. The only exception would be if the grotto or caving club knows them or has installed them itself. Equipment found in a cave is simply too risky. The danger of its being rotten or unsafe for any use is far too great.

3

Cave Sciences

Speleology, or the study of caves, is actually a subject made up of several separate but interrelated disciplines. These include geology, biology, hydrology, chemistry, and meteorology. Most tyros, soon after they start caving, develop an interest in how caves are formed and what kind of animal life they contain. This chapter will serve as a brief summary of these aspects of cave science. For those interested in going more deeply into the subject, an excellent and highly readable introductory text is: *Speleology, the Study of Caves,* by G. Moore and B. Nicholas, prepared in cooperation with the National Speleological Society. The material presented in this chapter is largely derived from that book. Also, an excellent and easy to read book about cave biology is *The Life of the Cave,* by Mohr and Poulson.

A cave is a naturally formed void located beneath the surface of the earth. By definition, it must have passages or rooms large enough to admit a human being, and by popular definition, must be long enough so that a caver can get out of the twilight zone into the zone of total darkness.

SPELEOGENESIS

How Caves Were Formed

Just exactly how caves were formed, is a subject that was not well understood until about thirty or forty years ago. Until that time, it was supposed that underground grottos were made by streams in much the same way that valleys and canyons are formed by streams above ground. However, there is now clear evidence that this stream action played only a very minor role in the formation of caves. In fact, it is now almost certain that extremely slow moving water, operating on a time scale of millions of years without the sounds, either of rushing water or heavy machinery that we might commonly associate with the digging of underground tunnels, is responsible for the beginnings of caves.

Not to get ahead of the story, however, it would be well first to understand that by far the vast majority of caves occur in limestone. Over the eons, limestone beds were built up layer by layer in a kind of massive burial ground for shellfish and other debris at the bottom of ancient shallow seas. Limestone is classified as sedimentary rock (the other types of rock are igneous—formed by molten rock cooling, and metamorphic, caused by rocks being changed by heat, pressure, or chemical action). Just as may be imagined, the beds of limestone were deposited unevenly with many layers of different types of rock, sand, or silt intermixed with the layers of limestone. The depositing process was often much like a snow fall with flakes of small limestone particles gently falling into layers and collecting in drifts, depending on the temperature, currents, and chemistry of the sea.

Over the ages, these limestone beds were shifted around and uplifted by massive action, like volcanoes and earthquakes. This upheaval caused vertical cracks in the beds, horizontal cracks being already present from the layering process, itself. Rains, floods, the various ice ages, and differing water levels, all added to the unending development that took place slowly during these geologic ages of the earth.

Many of the limestone beds in which caves occur are 250 to 350 million years old. However, the caves are much younger, usually less than 10 million years old.

Caves are formed in limestone (or its metamorphic version, marble) because the chief component of this rock is calcium carbonate or calcite, which is easily eaten away by carbonic acid. Carbonic acid occurs rather commonly in nature in a dilute

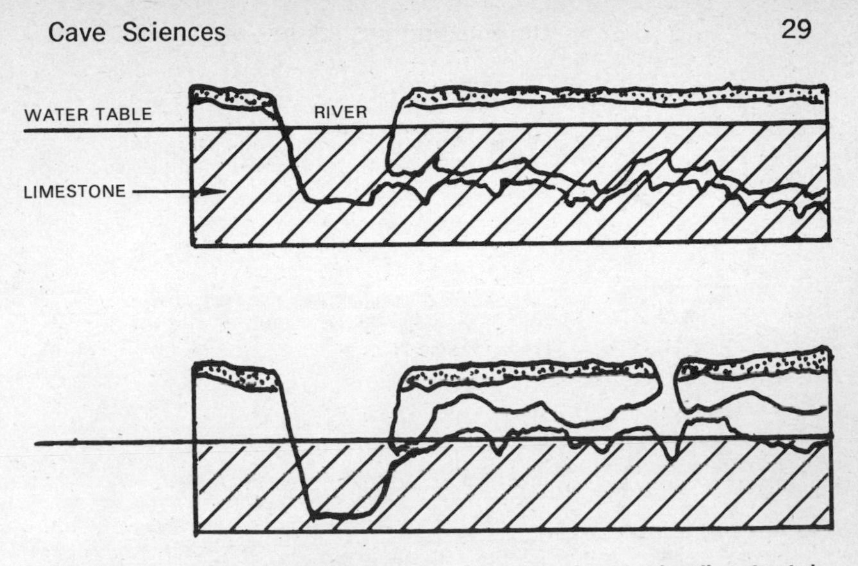

Fig. 3-1. Water Table—Cave Relationship. Caves form in water heavily saturated with carbonic acid just below the water table. Later, water drains from the cave as the water table goes down.

form. Even at full strength it is a very weak acid, but it was in its dilute form that it managed to slowly dissolve the limestone of these ancient beds and form caves. Carbonic acid is made when carbon dioxide reacts with water, carbon dioxide being produced rather abundantly by decaying plants, fires, and animal breathing. Chemically, carbonic acid is H_2CO_3. Surface water becomes charged with carbonic acid as it filters down through the soil into the cracks in the beds of limestone on its way to the water table (a layer of the earth saturated with water).

It is now generally agreed by speleologists that caves were formed by slowly moving water in the zone immediately *below* the water table. This water moves very slowly. In fact, it has been estimated that the rate of horizontal movement in a saturated zone can be ten feet or less per year. The exact shape and size of passages formed by this soluble water depends on the shape and size of the limestone beds—whether primarily horizontal or tilted up vertically by uplifting forces—and by the cracks and fractures in the limestone both of a horizontal and vertical nature. The resulting cave passages usually form some type of network so that a map of a cave often looks like a map of a city with many intersecting streets.

Development of caves immediately below the surface of the water table comes about very slowly at first, but then certain of

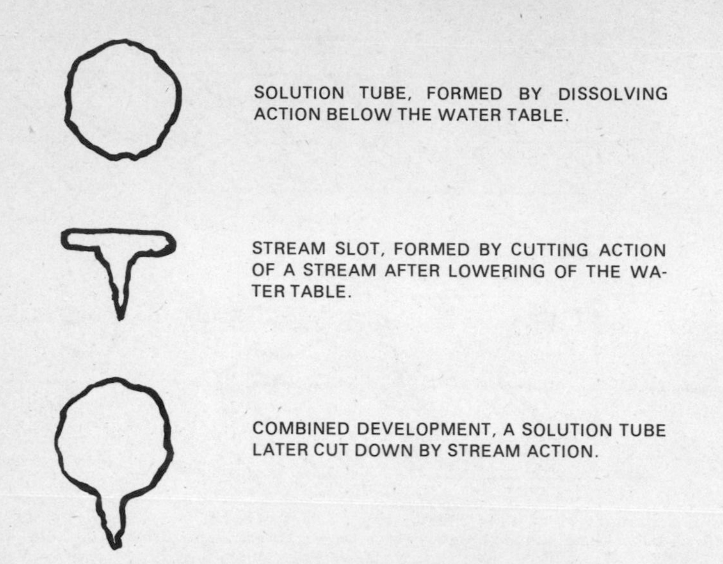

Figure 3-2. Typical Cave Passage Development.

the channels get larger and tend to take in more water and grow even faster. Eventually, the larger channels are able to carry all the available water and the others are abandoned. These passages then develop at a much faster rate and become the master passages of the cave. It appears that this solution activity in limestone keeps on going until the water table is lowered to a point where the passage is no longer flooded or when an entrance to the cave develops which allows the outside air to destroy the required carbon dioxide pressure needed for solution.

In the actual development of a cave, it may go through countless cycles of flooding and flushing, alternating with periods of complete dryness. Many cave passages clearly show this multicycle development. When a surface stream finds its way into a cave passage, small or large, it acts like a chainsaw because its fast flowing water is filled with gravel, sand, and rocks. This action is called corrasion, as opposed to the solution activity of corrosion by the carbonic acid, because it chips away at the floor of the passage (see Figure 3-2). However, as already stated, it would be a mistake to think that this corrasion action has been very important in the development of a typical cave. On the contrary, when a surface stream temporarily moves through an existing cave passage, it very clearly leaves evidence

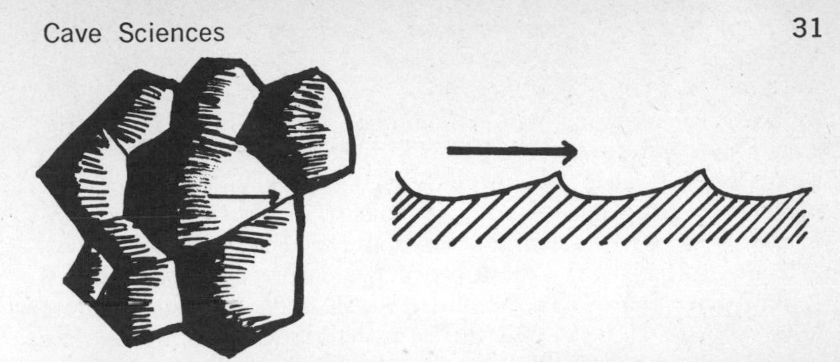

Figure 3-3. Scallops. Oval shaped solution hollows in a cave wall show presence and direction of the stream that caused them. Absence of scallops in most cave passages indicates that stream action played but a small part in the development of caves.

of its presence, as shown by small dissolved hollows, called scallops, found usually near the floor of passages, which indicate both the presence and direction of the flow. The steep slopes on a scallop are the upstream side and the gentle slopes are the downstream side. However, since scallops are not present in most cave passages, the evidence is clear that stream action has played only a small part in most cave development.
Fig. 3-3

Unlike the largely horizontal passages found in most caves, one type of cave room which usually is developed quite late in the history of the cave is called the domepit. These are usually developed after the water table has been lowered from the main cave passages and some entrance has allowed outside air to rush into the cave. During this stage, surface sewers may develop domes high above the cave passage and in some cases even extend the dome into a pit below the floor of the passage. These dome pits are characterized by vertical grooves on their walls, unlike the smooth walls found in other parts of the cave, caused by the surface water coursing swiftly down the walls.

Replacement Solution

Another mechanism of cave solution, only recently recognized and described (S. Egemeier, Ph.D. Thesis, Stanford University), is called replacement solution. Although this process has so far only been studied in caves with warm waters from thermal springs, it may operate in non-thermal caves as well. In the caves studied, groundwater rich in hydrogen sulfide feeds thermal springs in air filled (non-flooded) passages. Oxygen from the cave air oxidizes some of the hydrogen sulfide in the cave streams producing sulfuric acid which reacts with calcium carbonate (limestone) in the stream bed and dissolves it. Much of the hydrogen sulfide in the cave stream escapes into the cave air. Some of it in the air redissolves in droplets on the cave walls. Here it similarly is oxidized by the cave air, producing sulfur and sulfuric acid. This acid then attacks the limestone of the cave walls and converts it into gypsum. (Thus the limestone is *replaced* by gypsum.) Gradually, a coating of gypsum forms on the walls, and when it is one to two feet thick, the heavier pieces fall to the floor. Since gypsum is more readily dissolved by water than the original limestone, these pieces are rapidly dissolved by the cave streams. This solution process was actually witnessed in the thermal caves studied, a matter of only days or weeks being required for gypsum to be dissolved.

With this replacement solution now identified, the question is, have certain caves contained warm springs rich in hydrogen sulfide at some period in their long history, so that their development could have been aided by this much more rapid type of solution? If this is possible, then perhaps the well accepted theories of slow cave development just below the water table could come into question. Clearly, considerable study with this new theory in mind is required.

Cave Temperature

The temperature of limestone caves is remarkably stable during all seasons of the year, regardless of the outside temperature. In fact, a cave's temperature usually stays to within a degree or two of the average yearly temperature for its particular area. In the U.S., Canada and Northern Mexico, this ranges between 35° and 70°F, depending on the latitude and the altitude of the cave. So constant is the temperature in caves, they have been used for storage of food and other perishables for thousands of years.

As for humidity, except right near the entrance, the air in most caves is completely saturated with water vapor, which means the relative humidity is 100%. However, in certain caves in the southern parts of the U.S., humidity is less than 100%, and dust becomes a real problem.

Development of Speleothems

After a cave has been developed by solution, an almost opposite process takes place in developing the beautiful formations or

(Photo by Charlie and Jo Larson.)

The Big Column in Whipple Cave, Nevada. Note caver at base for scale.

speleothems. Speleothems are made up principally of calcite, the same mineral that was dissolved by acid charged water from limestone when the cave was being formed. When water reaches the interior of the cave, it contains a quantity of calcite in solution. This calcite is released as the carbon dioxide in the water escapes, causing the chemical change which is the reverse of that by which the limestone was dissolved in the first place. This results in the deposition of calcite on the cave ceiling, walls, and floor, which takes several different forms including stalactites, stalagmites, flowstone, canopies, rimstone dams, and cave pearls. Speleothems develop only rarely deep inside caves, because calcite deposition can only occur in well ventilated parts of the cave, depending as it does on release of carbon dioxide, not evaporation of water.

Stalactites hang down from the ceiling. They are actually tubular in construction, although in their larger versions this may not be apparent at first glance. They are built up as a drop of water coming from a crack in the ceiling loses its carbon dioxide, thereby depositing calcite in a tubular structure centered on the crack. Bit by bit, this grows into a tube, hanging down vertically. The growth of a stalactite is variable, but it has been determined to range from as little as 1/100th of an inch to 1/10th of an inch a year.

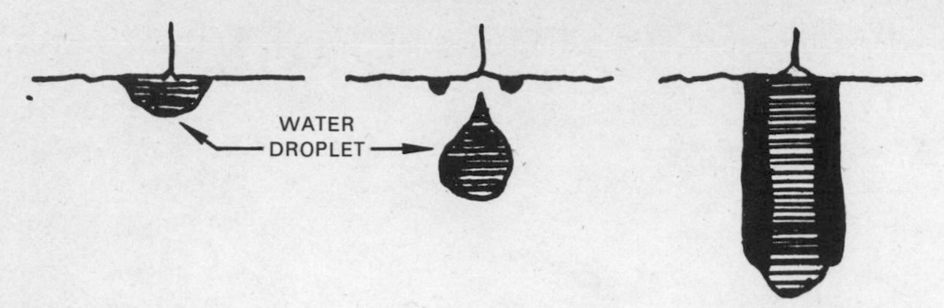

Figure 3-4. Formation of a soda straw stalactite. Water seeping from crack leaves calcite at perimeter of drop, which builds into the hollow straw configuration.

Stalactites can reach immense size and eventually can even grow to meet their counterparts, called stalagmites, which are rising from the floor to form what is called a column. Stalagmites are formed by the splashing action of water as it drips from the end of a stalactite. Often it still has some free calcite in it as it strikes the floor. In some caves, stalagmites can grow to more than fifty feet tall and thirty feet in diameter.

Sometimes water flows down the walls of a cave in a sheet rather than in drops and forms a speleothem called flowstone, one type often known figuratively as angel's wings. On occasion, these flowstone speleothems may be deposited over clay or gravel, which itself is washed out at a later time leaving an impressive canopy. Similarly, rimstone dams are formed in the floor of caves. The hypothesis for the formation of rimstone dams, which are relatively common in caves, is that as the water flows over the lip of the dam it is agitated slightly, causing the

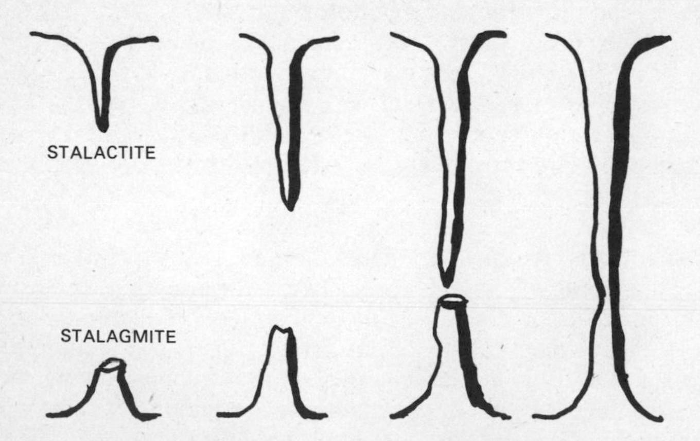

Figure 3-5. Stalactite hanging from the ceiling grows to meet a stalagmite rising from the floor to form a column.

release of carbon dioxide. This occurs more at the edges, which accounts for the development of the wall of the dam.

Cave pearls are another interesting formation made by deposition of calcite. These are formed much like the common ocean pearl around a small grain of sand.

Helectites, Shields, and Cave Flowers

Another type of speleothem found in caves are those deposited by seeping water. Probably the best known are helectites which seem to grow out from the wall in a manner completely defying the laws of gravity. They are about ¼ inch in diameter and often can reach a length of several inches. Most of them are of clear crystal composition although occasionally they may be white or even flecked with color. Apparently, they have a small central canal through which water is forced under hydrostatic pressure, but the flow is so slow that no drop of water is formed on the end which could allow gravity to have an effect on its shape. For this reason it grows in a completely helter skelter fashion. Helectites also occur of aragonite, as well as calcite, and usually look like very delicate needles.

Shields are another speleothem caused by seeping water. Quite rare in caves, shields are semicircular sheets which attach

(Photo by Charlie and Jo Larson.)

Helectites, a rare speleothem in any case, are shown here in an even rarer bush form.

themselves along the straightest edge to a ceiling, wall, or floor of a cave and project outward at different angles into the chamber. Cave coral, found quite commonly in caves, spherical stalactites, and cave flowers (often formed of gypsum), are the other principal types of speleothems formed by seeping water.

Standing Water Deposits

Cave bubbles and dog tooth spar are another of the curious speleothems found in caves. These are formed, along with cave rafts, either on the surface or under water in cave pools or lakes.

LAVA TUBES

Besides limestone caves, the other main type of cave is the lava tube. Lava tubes are formed when lava in a molten state is flowing from a volcano. As a tongue of lava flows down hill, its top surface and any exposed sides are cooled by the air and harden. But just a few feet inside, the molten flow is still in a liquid state and it moves forward until it runs out at the bottom of the tongue, or hardens to form an end plug if the ground

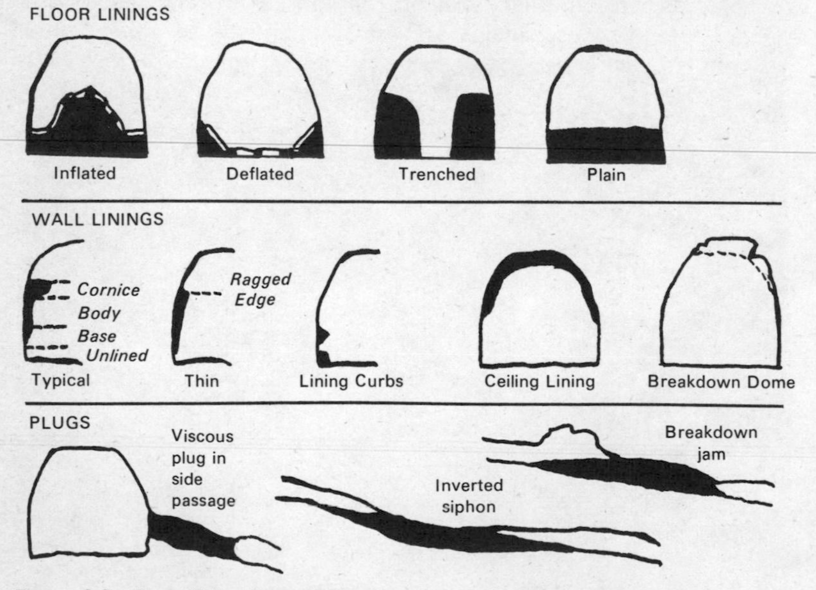

Figure 3-6. Lava Tube Passage Modification. Later lava flows and breakdown change both the main passages and side passages.

levels out. Recent evidence from studies of active volcanoes indicate that most of the movement of lava, even that of new flows in existing lava beds, is actually underground rather than on the surface. Thus lava tubes in areas of active volcanoes may be reformed or modified several times. Many lava tubes show clear evidence that several flows have moved down the same channel as shown dramatically in Figure 3-6.

Ceiling collapse leaving skylights in passages and breakdown on the floor are very common in lava caves. Speleothems, although formed by entirely different processes, such as soda straws, stalagmites, colums, and helectites are also found in lava tubes.

(Photo by Charlie and Jo Larson.)

Ice formations in a western lava cave. This is year around, not seasonal ice.

It would be a mistake to think of lava caves as only a few hundred feet long and more or less straight without many features. Actually lava cave systems can extend for many miles, with pits, crisscrossing tunnels, and maze-like complexity.

One unique feature of lava caves is the ice found in certain of them all year long. Because lava tubes are closer to the surface, they tend to follow the outside temperature much more closely than the very stable temperatured limestone caves. Thus, in lava tube with a small entrance and bottle shaped room below, dense air cold enough to freeze water is sometimes entrapped during the winter months. When this happens, the room can freeze any water that drips or flows into it and often an entire floor of good size will remain frozen even during the summer months. A cave visited in Washington during August of 1972 and called Ice Rink cave, had a room with a thirty by forty foot ice floor. The temperature in this room was exactly 0°C by actual measurement. Nearby rooms had a temperature of 4°C (39.2°F). One wall had a spectacular ice column formed by water dripping slowly down from the ceiling, that closely resembled this type of speleothem in limestone.

Other Non-Limestone Caves

Besides lava tubes, caves are sometimes formed in other non-limestone rock by the action of ocean waves. These are called sea caves, and occur mainly along the Pacific coast. They sometimes are also found in odd places like high above the water in the hills surrounding Great Salt Lake. These were formed by wave action in prehistoric times, when the lake was at a much higher level.

Sandstone caves occur in this soft rock, usually having been formed by the action of surface water and wind. They are normally very shallow and are commonly referred to as shelter caves. They are interesting, however, to the archaeologist, because they were indeed used for shelter by Indians, whose wall paintings (pictographs) and artifacts have been discovered in many of them.

Talus or fissure caves are not true caves at all, but are sometimes called that by local usage. They are formed by stream action washing away gravel or soft rock between boulders leaving a void, or by the fracturing of large rocks by ancient earth movements. This type of cave is not really altogether safe, because the rocks may not be as stable as they look at first glance. Hence caution should be exercised when checking them out.

LIFE IN THE CAVE

From the standpoint of ecology, a cave basically has two areas or life zones. These are the threshold or twilight zone, and the area of total darkness. Quite a few animals, particularly insects and smaller mammals like the rat, raccoon, and porcupine, use this threshold at least during part of the year, as their home. On the other hand, the zone of total darkness is quite inhospitable to life and supports only very specialized types of fauna. A novice must learn to look very carefully for cave life, because it can

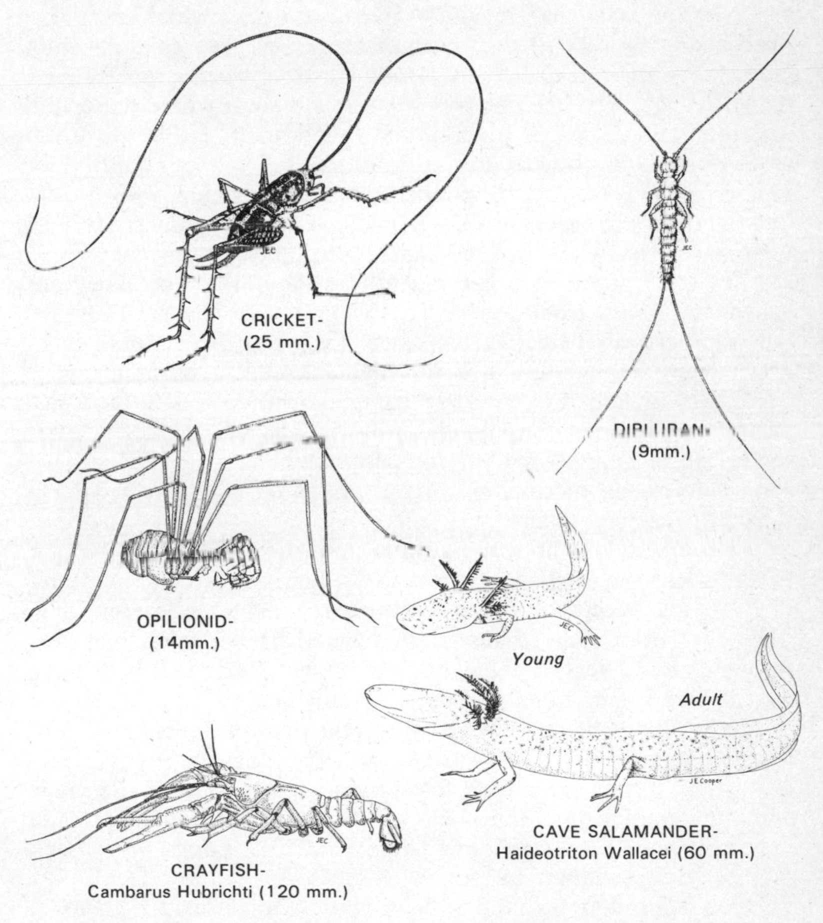

Figure 3-7. Cave Life

easily be missed unless pointed out by an experienced observer and it can also be easily destroyed by a careless step.

Cave animals are divided into three general classes, depending on the degree to which they are adapted to a cave environment. The first group is the one that lives above ground, but may use the twilight zone of the cave for protection from predators or the weather. This group is called *trogloxenes,* from the Greek words *troglos* (cave) and *xenos* (guest). It includes rats, porcupines, bears, skunks, raccoons, moths, mosquitoes and other insects.

In the dark zone, two different groups live. The first is a group of animals that regularly live there, but which can also live outside as well, so they are not classed as true cave dwelling animals. These are called *troglophiles,* from *troglos* (cave) and *phileo* (love), who live in caves because apparently they prefer the moist darkness. Some troglophiles spend their entire life cycle underground, but others of the same species live as comfortably in dark moist places above ground. This group includes some cave salamanders, crustaceans, spiders, millipedes plus insects such as springtails, flies, gnats, beetles, and mosquitoes.

The third group lives permanently in the dark zone. They are called *troglobites* from *troglos* (cave) and *bios* (life). They are the true cave dwellers that are born, live, and die in total darkness. They commonly have no pigment and have either very small eyes or none at all. They include blind cave fish, cave salamanders, isopods, amphipods, shrimp, crayfish, tiny clams, snails, slugs, and millipedes. Besides these animals, caves have several forms of microscopic organisms and bacteria which form a part of the food chain.

One of the best known creatures that use caves part time are bats. These small flying mammals, which resemble, but aren't really flying mice, use caves for hibernation in the winter time. There are even some caves in the Americas in which bats live throughout the year, leaving only to collect insects each evening for food, and some caves that they use only in summer as nurseries.

Bats, although the subject of a great deal of superstition and fear, are for the most part quite harmless. They have a remarkable sonar system which enables them to fly in total darkness without bumping into the walls of the cave or each other, and they also have a strong homing instinct which brings them back to their caves without fail.

One important precaution with bats. Don't disturb them in any way, particularly when they are hibernating or caring for their young. Leave the area quickly and quietly.

NEVER COLLECT

One of the strongest rules in caving is never collect any cave life except under the direct supervision of an experienced biologist. Observation is encouraged, but collection is prohibited! The very delicate chain of life in a cave is all too easily disturbed by a caver indiscriminately collecting rare forms of cave life under the mistaken notion that he is contributing to science.

Even just visiting a cave can change the ecology. The most damaging example of this is the partial destruction of the irreplaceable prehistoric cave paintings in Lascaux Cave in France, by simply opening them to human visitors. These remarkable examples of man's impulse to create art even under the most difficult conditions, were first discovered in 1940 by some boys looking for their dog. The paintings they discovered have now been dated as at least 20,000 years old. However, in the thirty or so years since the original discovery, some type of organism apparently introduced by the presence of people and lights have made the paintings deteriorate rapidly, causing the caves to be closed to the public until a suitable means of preservation can be found.

4

Getting Started

One of the easiest ways to find out if caving holds any real interest is to visit some commercial caves. (See Chapter 11 for guide books to commercial caves throughout the U.S.) Commercial caves afford an excellent introduction, both to caving and to cave conservation. In many instances, they are even more beautiful and better preserved than the average wild cave. The reason for this is obvious: the owner selected a cave with beautiful speleothems (formations) in the first place, and now has a strong economic interest in displaying this beauty in the most attractive way possible. Besides, he or a caretaker are always at the cave which is why there is seldom any vandalism in commercial caves.

From the standpoint of the novice, a visit to a commercial cave can tell him if he is likely to suffer from claustrophobia or from a fear of darkness. In many caves, guides will turn off the lights for a minute to let visitors experience total darkness. If this or claustrophobia bothers one, he should think twice about caving. Or perhaps he will find he is even more attracted to the beauty of caves and thus may become more interested in working with organized cavers for cave conservation, rather than becoming a full fledged caver. In either case, visits to commercial caves can help focus one's interest on the underground wilderness.

A word about courtesy when visiting a commercial cave. Do the owner a favor and don't stray off the commercial tour. It can be just as dangerous caving alone in a commercial cave as in a wild one. Save those spelunking instincts for caving with an organized group in a wild cave.

On the other hand, the record of cooperation between chapters (grottos) of the National Speleological Society and other caving clubs with commercial cave owners is long and growing all the time. There are countless cases where cavers, after a proper introduction, have helped discover, map, and even develop whole new areas of well-known commercial caves.

GROUP CAVING

Organized Group

If the caving aspirant finds that his preliminary paths of inquiry still lead underground, then he should know that the only safe way to start caving is to find some experienced cavers and go on a trip with them. Seriously interested neophytes are invariably welcomed by caving clubs, since training and conservation are among their primary goals. Finding experienced cavers shouldn't be too hard in most areas.

The National Speleological Society

A good place to start is to get the address of the nearest grotto of the National Speleological Society from its new (since 1971) centralized headquarters. Simply write NSS, Cave Avenue, Huntsville, Alabama 35810. Please enclose a self-addressed envelope.

A neophyte caver will find that if he's at all interested in caving, the National Speleological Society is where most of it is taking place. Besides local chapters and activities in the U. S. and Canada, the Society holds regional conventions in local areas topped off by its annual convention. This affair is regularly attended by 500 to 600 enthusiastic cavers and speleologists. It's a chance to meet everybody who's anybody in caving, hear about the latest techniques and scientific theories, and exchange stories with the people who really love caving.

Rotated each year among the best caving areas, it combines field trips with meetings and fellowship. Most cavers camp at NSS conventions, but plently of overnight accommodations are also arranged for by the host grotto, sometimes in college dormitories at the increasing number of university hosted conventions, or in motels.

Other Caving Clubs

Another way to locate groups that sponsor organized caving trips is to check the outing clubs at local colleges, universities, local sporting associations, athletic clubs, and similar organizations. In some cases, the college groups will be student grottos of the National Speleological Society. In other cases, they may be a section of a recreation group involved with several outdoor activities such as canoeing, mountain climbing, biking, or backpacking.

A third place to find cavers is the Sierra Club Chapters around the country. Some of these will have caving activities, particularly if they have a rock climbing section. Again, many of these cooperate and share trips with nearby NSS groups. Information about Sierra Club Chapters is available from the Sierra Club, 1050 Mills Towers, San Francisco, California 94104. Again, please enclose a self-addressed envelope.

Finally, owners of wild caves and commercial cave operators often know of local caving clubs that have visited their area and can refer inquiries to them.

Importance of Group Caving

Just as caving with a group is the best and safest way to get started, there is no doubt that the worst way is for a novice to collect some candles and a ball of string and go off caving by himself or with another neophyte. Caving alone is suicide. It's like someone walking the plank voluntarily, with no one pushing him, and discovering when he hits the water that he forgot his water wings.

Caving with a group offers three key benefits. They know how to cave (and can teach a newcomer), they have equipment like ropes and ladders that the individual caver can't always afford, and they know where the caves are.

The fact is, grottos and other clubs not only know where the caves are, they know what equipment is needed, how dangerous

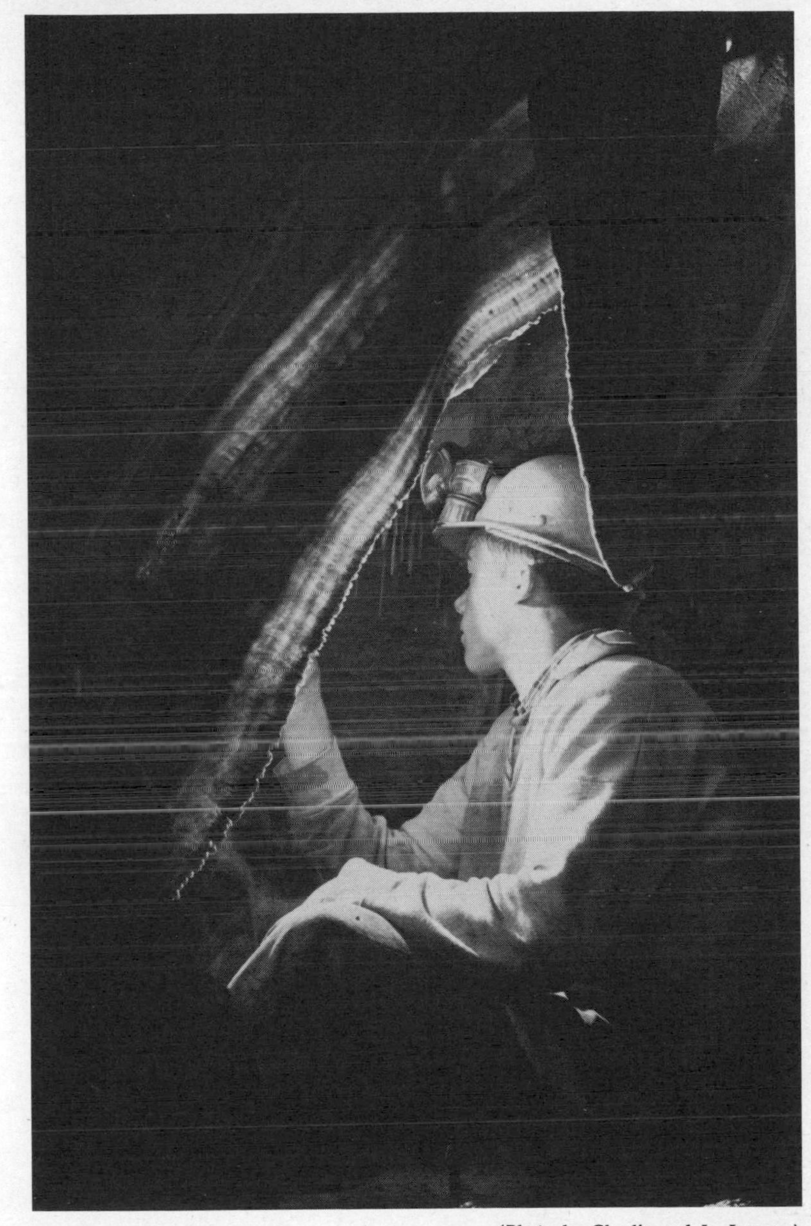

(Photo by Charlie and Jo Larson.)

Drapery or bacon rind in a nothern California cave.

the caving operations will be, and just about anything else one would want to know as a beginning caver.

Fellowship is also a benefit of club membership. In a group, one will find others fully as interested and enthusiastic about caves as anyone.

Historically, as caving developed, clubs sprung up to pool skill, knowledge, and equipment and to standardize on techniques. Today, it's very rare to meet a serious caver who doesn't cave with a group, be it an NSS grotto or a local caving club.

WHO GOES CAVING?

To the uninitiated, it may seem that caves are only dark, dangerous holes in the ground in which it is altogether too easy to get lost. But to thousands of dedicated people, they offer a unique challenge, both as a sport and a science. The ranks of the caving fraternity are mostly made up of the eighteen to thirty-five age group of both sexes, but there are also many families with teenagers. Even small children go caving in beginner caves. Beyond the 18 to 35 group are a sizable number that have become seriously interested in one or more special aspects of caving as an avocation (exploration, techniques, photography, cave sciences, conservation, society administration). This group tends to continue for many years and to provide a firm base for the educational and conservation goals of organized caving.

Why Go Caving?

People go caving for many reasons. Some go for the sense of achievement—to overcome potential hazards with the proper technique and equipment. Some want the pure adventure—the thrill of exploration into the unknown, of finding a new cave or a new section of a known cave, and being the first to explore it. Others enjoy the sense of companionship and teamwork that cave exploration demands. But whatever the initial reason for going, cavers soon join together under the common bond of shared experience underground.

Caving is similar to mountain climbing, except that a climber can usually see his ultimate goal above, the top of the mountain. A caver on the other hand, can't see the end of the cave when he starts out. At least he hopes he can't. He pushes on in the hope that the end is *not* just around the next corner, but many miles further ahead. In the language of the caver, he hopes that it goes.

Fellowship of Caving

Fellowship is another important ingredient of caving and cave exploration. However, fellowship actually means more than just having a good time at a meeting or some social get-together. In caving (as in mountaineering or scuba diving), one comes to depend on the other members of the team for help and assistance and in some instances for one's life itself. This creates a bond between fellow cavers that is sometimes difficult for non-cavers to grasp at first. Cavers are not sensationalists and seldom like to discuss their underground activities with non-cavers for fear of overglamorizing certain aspects without conveying the proper feeling for safety and conservation. But neophyte cavers invariably find that the NSS grottos or other caving clubs will welcome any person seriously committed to conservation and interested in pursuing caving or speleology.

The Caving Season

One appeal of caving for many people is that it is pretty much a year around activity. About the only limiting factor is severe cold, snow, or heavy rain. Any of these can make it difficult to reach a cave entrance from the nearest all weather road. Once inside a cave, however, temperatures (which range from 40° to 70°F, depending on latitude) are nearly constant within a few degrees all year long, so they will often be refreshingly cool in the summer or relatively warmer in the winter.

5

Personal Caving Equipment

It used to be said that all a caver needed for a trip underground was some old clothes, a hard hat, and a carbide lamp. While it's true that many cavers give the impression that they never wear anything but old clothes (and discards from a surplus store at that), their personal equipment is really chosen with great care, each piece serving a definite purpose and selected accordingly. If the average caver tends to look a little rumpled and muddy, it's only because the rugged cave environment makes all articles of clothing, new or old, take on the patina of wear after their first use in a cave. One thing is for sure—cave mud penetrates all the way to the skin, which makes any clothes, no matter how many layers down, usable only for caving after their first trip into a cave. So be prepared to have one set of clothes for caving and another set of clothes for more normal activities.

To get started, all a caver actually needs is a borrowed hard hat and light, some old clothes, and a good pair of hiking boots. The boots preferably should have lug soles which are also useful for hiking and rock climbing if caving proves a washout after the

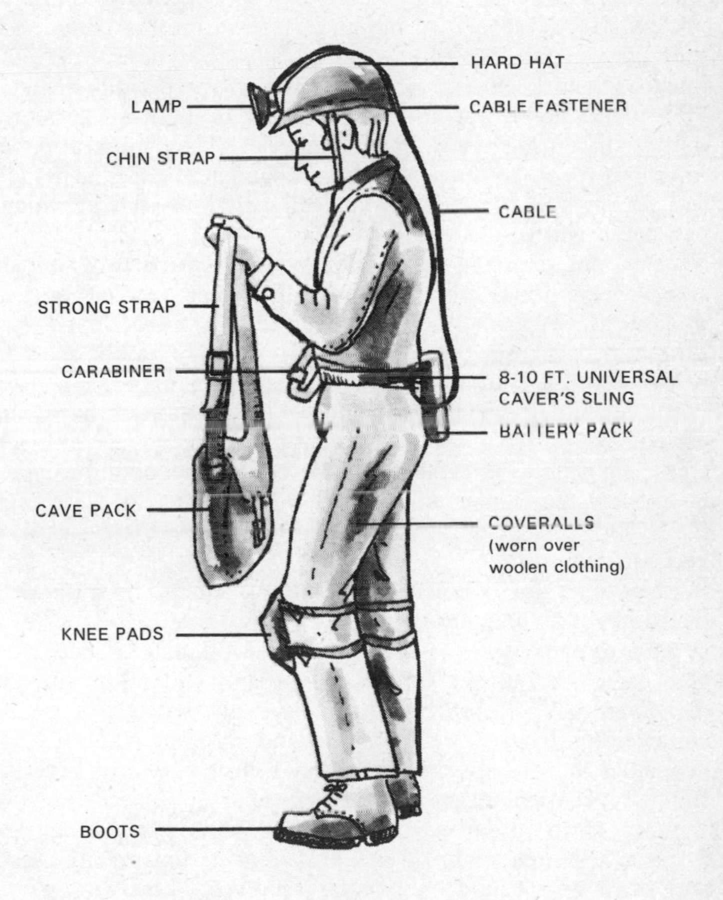

Figure 5-1. A Fully Equipped Caver.

first few tries. Besides the borrowed light, the first-time caver should plan to bring his own flashlight, candles, and matches in a waterproof container (for his other two sources of light), plus extra carbide or batteries for twice the expected stay on the first trip. Taken together, the hard hat and carbide or electric light represent an expenditure of about $15, for those who want to get their own.

CLOTHING

Choosing the proper clothing for caving depends to some extent on the type of cave trip, but there are still some common items regardless of the type of cave. Basically, the job that cave clothes must do is to keep a caver warm enough for periods of inactivity (fairly common), yet not be too bulky or tight so as to restrict movement during crawling or climbing. They must also be free of floppy pockets, loops, or belts that catch on sharp rocks or projections. In caving, it is a dead certainty that anything that can catch will catch!

For underclothing on anything but a one or two hour excursion, cavers today almost universally wear woolen or thermal underwear. Woolen longjohns over regular underwear are much preferred because they retain heat better than thermals which are usually all cotton. In some cases string or net underwear is worn next to the skin for the uppers. Recent tests by mountaineering groups have shown that wet cotton clothes are actually worse than no clothes at all under cold, wet conditions, so wool is strongly recommended to all cavers. Cave temperatures range from 40°F or less in the North and Canada to 70°F in the South and Mexico.

Not only are woolen longjohns beneficial because of their heat-retaining characteristics, but in any sport involving exertion, several thinner layers are always better than one because they give more air barriers for insulation and still allow freedom of movement. Wool absorbs water very well (which in a cave could come either from perspiration or underground water), and still keeps one warm. For caving in very wet or very cold caves, a wet suit is recommended, as discussed in chapter 7.

Over the wool underwear, a pair of pants and shirt are usually worn. Although blue jeans are almost universally used by American and Canadian cavers, relatively loose fitting woolen trousers of either flannel or worsted, which give better protection and warmth, are recommended instead. The cotton denim of blue

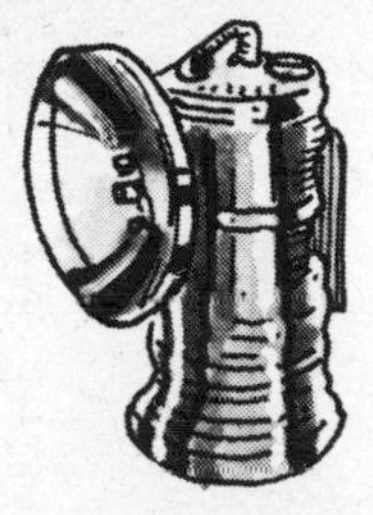

CARBIDE LAMP

ELECTRIC LAMP AND
BATTERY PACK

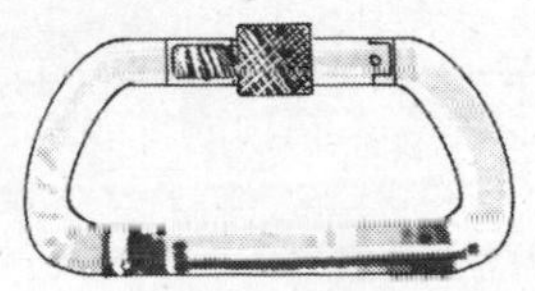

LOCKING CARABINER

NARROW WELT,
LUG SOLED BOOTS

HARD HAT, METAL OR
FIBREGLAS TYPE

Figure 5-2. Typical Personal Equipment.

jeans is really too thin and hard to offer the knees much protection if crawling is to be encountered, as it almost always is. Also, blue jeans are too tight in the hips and legs for many people to allow comfortable movement for the climbing and crawling common in caving. Finally, cotton just isn't as good as wool when it comes to keeping a caver warm in the wet, muddy underground.

As with the pants, a caving shirt should be relatively loose fitting. Many people wear flat weave cotton shirts, but here again a woolen shirt is preferable for warmth. Be sure it has long enough tails to stay tucked in when bending over, reaching up, or twisting.

Over the pants and shirt, most cavers prefer a suit of coveralls (called a boilersuit in Britain). Coveralls, as the name implies, cover everything and keep other clothes in place. This is more essential than might be imagined when squeezing through a tight vertical slot or horizontal crawlway. While it is true that some cavers continue to wear a sweatshirt worn outside the pants and pair of blue jeans, most who regularly encounter any kind of tight passage have universally adopted the coverall. When shopping for a set of coveralls, try to find one with buttons rather than a zipper. It turns out that the zipper is usually the first thing to go in a set of coveralls (the seat is a close second) because of the mud and water in caves.

One important point: don't load the pockets of either the coveralls or pants with anything hard. Crawling on a tin of spare parts or a pocket knife can be quite painful. To prevent pants and coveralls from riding up on the legs, some cavers tuck them into their socks. This is reasonably effective. A better method is the gaiters of the British caver, though these have never caught on in this country. However, gaiters are sometimes available in surplus stores and might be worth trying if they can be found.

For cavers who expect to be sitting a lot during surveying, photographic trips, or similar projects, a good insulation for the seat of the pants is ⅛ inch Ensolite or equivalent (used for sleeping pads by backpackers). Cut a piece wide enough to extend from hip to hip, and long enough to go down about halfway or less to the knees. To act as a pocket, cut a piece of old sheeting or denim an inch or so larger than the Ensolite, and sew it inside the pants, slotting both it and the pad to fit down inside the legs. Leave the top of the pocket open, so the pad can come out when the pants are washed. (Our thanks to Luther Perry of the Southern California Grotto for this very practical idea.)

Extremities

For the feet, two pairs of wool socks are recommended, one light and one heavy, although some cavers prefer both to be heavy. Again, wool is preferable to cotton because it absorbs water and is warmer. For the hands, gloves are really an essential for almost any type of caving. The best types of gloves are leather-faced cloth gloves and plastic knobby-palmed gardening gloves, both of which tend to be a little warmer than plain cotton gloves. However, plain cotton gloves have the virtue of being very inexpensive and can be easily discarded if they don't recover after washing. Leather gloves, such as the thin gloves much in favor with sports car drivers, are also very good because they give a nice close feel when climbing and scrambling underground. Unfortunately, they are quite expensive and require a great deal more care to keep soft and pliable.

A note about nails—finger and toe, that is. Be sure finger and toenails are clipped on the short side before caving. A caver's hands get so muddy that long fingernails become reservoirs for what seems like acres of dirt. Besides, long fingernails get in the way and are easily broken. Long toenails are also a bother since they can be forced against the toe of the boot, which may be painful or at least irritating.

BOOTS

Obtaining the proper boots for caving is so easy today, that there is really no excuse for a caver being unprepared with improper footwear. Nearly every sports store, including the sports departments of most department stores, have narrow-welt lug-soled boots. One of the most common (and best) lug soles is called the Vibram sole, although there are several very similar newer brands on the market now.

Buying cheap boots is a very poor economy. Many cavers feel that because cave boots take such a beating they are reluctant to pay for an expensive pair of hiking or rock climbing boots. However, cheap boots can easily fall apart and cause injury to the feet. Proper footwear in caving is just as important as in hiking or mountaineering.

Caving boots should be ankle high (five to six inches) to permit good flexibility for climbing, traversing, and scrambling. The toe should be hard so it protects the toes and allows pushing off with the foot in tight crawls. Avoid tennis shoes, other types of soft shoes, or oxfords. Low shoes are very often left behind in the mud accompanied by a strange sucking sound as a caver tries to step forward.

Types of soles to avoid are smooth soles, particularly crepe soles which will slip on any wet surface and should never be used in caving. Similarly, nailed climbing boots are unsuitable for caving, though they may seem to make sense for muddy rock. The problem is that nailed boots can easily damage ladders, cave formations, and climbing rope. Generally cave boots will weigh somewhere between 2½ to 5 pounds. Avoid the very lightweight rock climbing boots without heels. A heel is needed both for comfort and to hook onto the rungs of caving cable ladders.

Vibram soles are ideal for caving, but watch for wear. When the lugs of the sole get too worn, as evidenced by rounded edges, Vibram soles are not really much better than smooth rubber soles which are themselves downright dangerous in a muddy cave. Vibram soles grip well on both wet and dry limestone, but they may slip on some muddy surfaces. If a great deal of mud is encountered, stomp the foot sharply to remove mud from between the lugs before attempting any climbing or traversing involving small footholds.

When trying on a pair of boots, wear the same two pair of woolen socks that will be worn underground. Watch particularly for snugness at the instep and be sure that the lacing of the boot holds the foot well back into the boot. With the laces tied normally, bang the toe of the boot on the floor. If the toes feel it, the boot is probably too small. As for lacing, the types that use holes rather than clips are better because clips can hang up in a tight spot or on a cable ladder.

The overall fit of caving boots should be somewhere between the extreme of very tight needed for rock climbing and the loose fit needed for hiking. In general, it would be better if the fit were too loose rather than too tight because of possible circulation problems in a cold cave.

One of the advantages of Vibram soles is that when worn down, they can be replaced at about half the cost of new boots at many local shoe repair shops. Be sure the cobbler understands that the final sole needs to be trimmed very close (either completely flush or even slightly undercut) to the sides of the boot. This is called narrow-welt construction and is vital for caving and climbing. It's well to point out to him the narrow-welt construction of the boots when they are first taken to him so that he will cut the new sole properly to maintain the narrow welt. Cost of resoling caving boots with Vibram soles will run between $10 and $15.

Boot Care

To care for caving boots, dry them slowly, preferably by stuffing them first with paper. Don't dry them in front of a fire or heater because this will cause the leather to crack. Be sure to dry them in a flat position on the soles, otherwise the sole may crack. When dry, it will be surprisingly easy to brush off loose mud. Treat the leather as any other good pair of shoes with polish, saddle soap, or suede conditioner, depending on the finish.

HARD HATS

No experienced caver ever sets foot in a cave without a hard hat. A hard hat protects the caver in two ways. Primarily it is to keep from banging the head when ducking under rocks and overhanging formations. But it also protects the caver from falling rocks. This ever present danger (although not as common as in rock climbing where helmets are seldom worn) is still an almost sure killer, since even a small stone, let alone a boulder, can reach lethal speed in a cave. And remember, a cave hard hat without a chin strap is next to useless.

To speak from personal experience on this point, on a recent (1972) cave trip, a 200-pound rock, apparently loosened by water or caver traffic, suddenly broke loose from its position in the cave entrance. It fell only 23 feet, but it moved with the deft precision of a well aimed billiard ball, managing to hit all four cavers positioned several feet apart at different points on the entrance slope as it richocheted down. In the case of two of them, it undoubtedly would have killed them if they hadn't been wearing hard hats. As it was, one suffered severe forehead injuries, *after* the rock had cracked his hard hat. The moral is: *Don't cave without a hard hat equipped with a snug chin strap.*

In the last couple of years, most cavers have standardized on either the plastic, metal, or fiberglass hard hats used by the construction and mining industries. When buying a hat, be sure it has been certified as meeting state or national safety standards for construction of hard hats.

A new type of hard hat recently used by some cavers is quite warm to wear, but may be the best type of all for protection against falling rocks. This is a high density linear polyethylene helmet worn by motorcyclists and racing drivers. After all, these drivers are engaged in a far more dangerous activity than caving, and a lot of research lies behind the design. This type helmet fits around the head much more snugly, which makes it very comfortable, but too warm for some without the headwinds of a 70 mph motorcycle ride. Certain types of ex-military helmets, both the steel pot or the helmet liner, itself, had some currency in previous years, but don't wear these. They are totally unsafe for caving use.

Equally important as the outer shell is the suspension. This should be a miner or safety suspension of the approved polyethylene construction in which all the inner edges of the suspension are rounded.

A caver's hard hat always has a chin strap. It never takes more than one trip back down a slope in order to retrieve a dropped hard hat to convince a caver that a chin strap is necessary. A caver's helmet with the extra weight of the carbide or electric lamp falls off the head all too easily. Hence the chin strap. Some cavers even install a lead weight on the rear of the helmet to counterbalance the lamp weight—something worth trying if one encounters such a problem.

Finally, on the front of a caver's hat is a lamp bracket. The most common type accepts the flat spade bracket of the carbide or electric lamp. Some older ones will also accept the wire bale found on earlier model carbide lamps.

HEAD LAMPS

It wasn't too long ago that the carbide lamp was standard for all cavers. However, in the last two or three years the electric lamp has achieved such a popularity that it is now used by about half the cavers in the National Speleological Society. Many cavers apparently feel that if they are going to cave seriously for periods of more than eight to ten hours, an electric lamp, despite its higher initial cost, is a better way to go. Needless to say, voicing this opinion will produce an instant argument from the carbide contingent (myself included!).

Carbide

Carbide lamps are quite simple in principle. They use small lumps (¼ inch) of carbide (calcium carbonate), onto which drops of water are released to produce acetylene gas. This gas

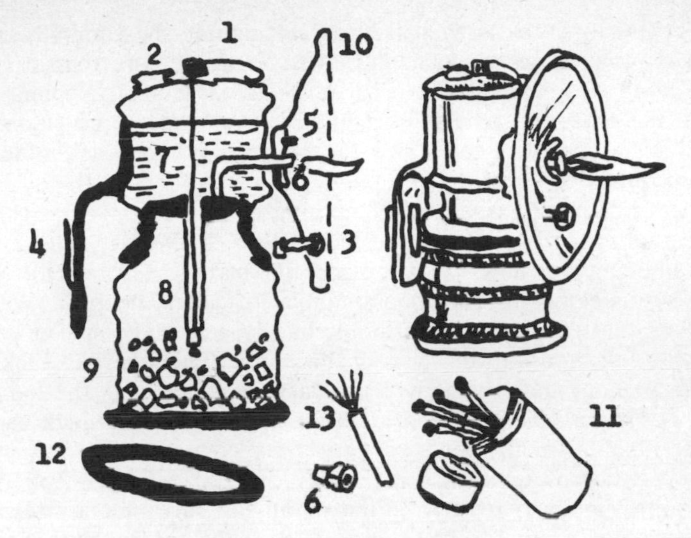

Figure 5-3. Carbide Lamp.

1 - CONTROL LEVER
2 - WATER FILLER CAP
3 - FLINT AND WHEEL
4 - SPRING CLIP
5 - NUT
6 - TIP
7 - WATER CONTAINER
8 - WATER LEVER
9 - BOTTOM WITH CARBIDE
10 - REFLECTOR
11 - MATCHES (in waterproof box)
12 - RUBBER WASHER
13 - TIP CLEANER

is directed up via a pipe to a tip with a small hole where the carbide mixes with air and can be ignited easily with a spark, match, or the flame of another carbide lamp. Up until now, all carbide lamps have been made of brass. Recently (late 1971), the major American manufacturer, Justrite, has brought out a plastic model. By all reports, this new lamp—unless modified or greatly improved—does not fill the bill and should be avoided. The beginning caver is well advised to seek out a brass lamp now and perhaps buy a spare before the old Justrites disappear. Several British lamps, notably the Premier, are very similar to the Justrite and have several interchangeable parts.

A carbide lamp consists of two main parts, a top with reflector, tip, and water container, and a bottom which contains the carbide. The lamp bottom screws into the top, a gas-tight fit being insured by a heavy rubber gasket between the edge of the bottom and the lip on the top.

In operation, one charges the lamp by filling the bottom about one-half to three-quarters full with carbide. Next, fill the top with water and adjust the flow by moving the water lever on the top one or two clicks. It's best to do this before reassembling the lamp and actually watch for a steady stream of droplets. To give the acetylene a head start, many cavers add some water (to put it crudely, they spit) into the bottom and then assemble the lamp quickly.

At this point gas should be coming from the tip, which can be detected easily by holding the lamp close to the face and feeling or smelling the flow. Next, place the hand over the entire reflector assembly to entrap a quantity of acetylene gas. Wait a few seconds, then draw the hand sharply across the striker wheel on the reflector with the heel of the hand. This operation takes a little practice but it's really quite easy once one gets the hang of it. If the gas has ignited, a loud pop will be heard, which can be deafening in small rooms or passages.

If it doesn't ignite, there can be several causes, each of which is quite simple to remedy. First of all, be sure that a stream of acetylene is coming from the tip. If it is not, the chances are that the tip is clogged and this can be easily corrected with a tip reamer or cleaner. Check, too, to be sure that the bottom is tightly screwed into the top. If it is not, a great deal of acetylene gas may be leaking out this way rather than travelling through the tube to the tip. This can be checked by holding an open flame from another carbide lamp to the bottom to see if it will burst into flame. If it does, the remedy is usually a simple tightening of the bottom. If this doesn't work, try turning the gasket upside down or wetting it before reassembling.

Be sure not to turn the lamp upside down or it may wet the felt through which the acetylene gas moves on its way to the tip. If this felt becomes wet, the lamp will become very balky in operation. The flame will alternately go from very short to very long. It may whistle shrilly, and then go out. The only solution is to remove the felt and replace it with a dry one. If no dry felt is at hand, try wringing out the wet one and drying it gently with the flame of another carbide lamp or candle.

Usually, one charge of carbide will last two to three hours. During this period, it will probably be necessary to increase the setting of the water lever several times. When doing this, keep in mind that it takes ten to twenty seconds for the lamp to respond to a change in water. Also, be careful not to get too much water in the bottom, because this will produce what is called flooding, a very messy situation, usually indicating the carbide is spent or nearly so. After about two to three hours, the lamp will run down and the flame will become lower and lower. Be sure to blow out the flame, before it gets so low that it clogs the tip with carbon. If this has happened, a tip reamer can be used to remove the carbon.

At this point it's time to "carbide up," as cavers call it. Depending on the length of the trip, one of the easiest ways to handle this refueling is to have five or six separate carbide bottoms (which come complete with a gasket and top) already full, and simply exchange the old one for the new one. When doing this, don't tighten the top on the bottom with the spent carbide too snugly. This is a safety precaution in case there is life left in the carbide and any acetylene generation is going on. By putting the top on the bottom just snugly any acetylene can leak out safely.

Some cavers don't go this route because of the added bulk and expense, so the other method is to carry along a plastic bag and empty the spent carbide into it. Extra carbide is then carried in a waterproof container, such as a plastic baby bottle. Leave the nipple upside down in the baby bottle to act as a gasket when tightening the plastic top.

When tapping the lamp bottom to remove the carbide to the plastic bag, be careful not to strike the threads or top of the lamp bottom because this can bend the threads and cause a leaky joint between top and bottom. If it's necessary to loosen up the carbide in the bottom, use a knife or stick or tap it on the bottom—not the top. Refill the bottom with another half to three-quarters load.

After recharging by changing a bottom or refilling, check the water supply at the same time. It is probable that the water will have been replenished at least once or twice already since water tends to run out more quickly than carbide. Be very careful when blowing away spent carbide from lamp parts or from the bottom. It's very dangerous to the eyes, particularly if it's rubbed in. Those who get a good exposure to carbide in their eyes should see a doctor immediately.

As indicated earlier, *never leave carbide, whether buried or not, in a cave.* Always carry out all spent carbide. And similarly, don't leave spent carbide around the entrance to a cave or anywhere on the owner's property. Spent carbide is very dangerous to livestock. Several cows have died from the thoughtless acts of cavers leaving spent carbide above ground.

Cleanliness is next to caviness as far as proper carbide lamp operation is concerned. Be sure to clean the lamp carefully, occasionally taking out the tip to be sure there is no accumulation of carbide in the gas tube. Similarly, check the condition of the flint, felt, and felt retaining clip, and replace if necessary. Polish the reflector, using soap and water. Avoid abrasive cleaners. Usually, felts are only good for two or three cave trips and then they should be discarded. It's possible to rinse out the accumulated carbide dust from a felt if necessary, but it's better to have five or six spare felts at all times since they only cost a few pennies each.

Spare parts for a carbide lamp are essential on each trip. Besides spare carbide, a spare parts kit should include a tip reamer (either the single needle type or the flexible wire type), two or three tips, gaskets, felts, flints, plus a complete striker wheel assembly, and felt retainer clips.

When using the lamp underground, adjust the flame for a one-half to one and one-half inch flame. A longer flame than this will give a great deal more light and seem comforting to the new caver but it will use up carbide much too quickly. A properly adjusted lamp will supply two and a half to three hours of light at a minimum with no trouble at all.

Another advantage of carbide lamps, besides their lower cost, is their open flame. Sometimes an open flame is useful in a cave for fusing rope ends, cooking, or heating water. The major disadvantage of a carbide lamp is that it blows out quite easily in windy passages and is next to useless when climbing in a waterfall. Its flame can also cause damage to climbing ropes when on belay, prusiking, or rappelling, though experienced cavers don't seem to have any problem directing their flame away from ropes.

Electric

Electric head lamps have greatly gained in popularity in recent years because of their improved long term reliability and life. One minor problem with electric caving is that it's almost impossible to buy a really satisfactory lamp without making some modifications. However, it seems that it must really be worth the effort because the number of electric cavers has increased tremendously.

Basically, an electric lamp consists of a lamp assembly and a separate battery pack. This battery pack is connected either with a fixed cord or a cord and plug to the lamp. It is usually positioned on the belt or hip or in some cases has been carried in a separate shoulder pack. Some cavers mount the cord inside the coveralls to avoid the most serious problem of the electric caver, that of developing a snagged battery cable. A hung up battery

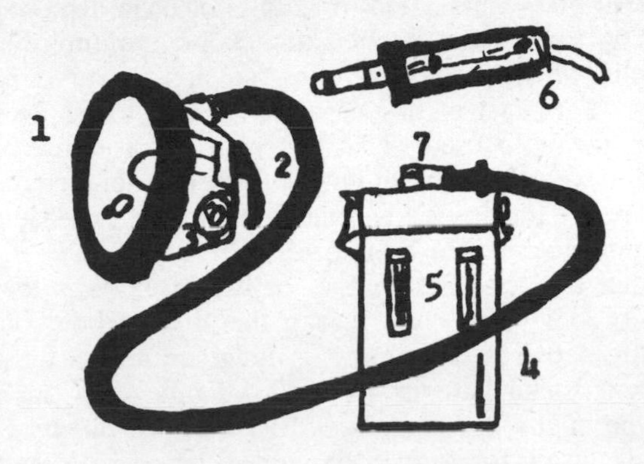

Figure 5-4. Electric Head Lamp.

1 - LAMP ASSEMBLY WITH TWO BULBS
2 - SPRING CLIP FOR HELMET
3 - SWITCH
4 - BATTERY PACK WITH NI-CAD BATTERIES
5 - SLOTS FOR BELT
6 - DETATCHABLE CABLE PLUG
7 - CABLE JACK

cable can be a real problem in a tight cave where it may be very hard to back up and free the cable from even a small protuberance.

Some of the simpler modifications that can be made to the standard head lamp available from the Justrite Corporation are as follows: either specify when buying or else order a replacement lens of honeycomb construction rather than the clear glass supplied with the lamp. This tends to diffuse the beam and make it more suitable for caving use. If anyone has ever caved with an electric caver who had clear glass in his lamp, he is familiar with the blinding effect caused by the electric lamp shining full force into his eyes. Secondly, most cavers replace the Justrite battery pack completely or put pieces of cardboard in the stock battery pack to hold the batteries tighter. Finally, it's a good idea to install screw-on connectors either at the lamp end or the battery pack end (or both), to relieve the "snagged cable" problem. Several other brands of electric lamps are seen in caves, including Mine Safety Appliance and Wheat Light. Some have a reflector with two bulbs, one of a much lower current drain for emergency, the other for regular use. Check the cavers in the local area for suppliers and any needed help on modifying a lamp assembly to match specific needs.

Those seriously interested in using electric lamps generally end up using some type of special nickel-cadmium cells. These are wet cells which through recharging can be used essentially indefinitely. They last eight to twelve hours on the average, depending on the bulb used. Ni-Cad cells, as they are commonly referred to, have a high initial cost, but in only one caving season they will pay for themselves as compared to the expense of standard flashlight (zinc carbon) batteries.

The other type of cell suitable for use in caves is the alkaline cell. This is somewhat heavier than the zinc carbon, but it lasts anywhere from five to ten times as long, or allows about six to eight hours of steady use in a cave. Alkaline cells are probably the only type of dry cell that should be used for caving. Mercury cells, widely used for transistor and other low current drain applications, are unsuitable for caving use because they're too sensitive to cold and they lose their voltage when the temperature drops. Zinc carbon cells wear out too soon. Any dry cell can have its storage life extended by storing in a refrigerator. Allow a couple of hours for recovery before putting them to use underground.

The advantage of nickel cadmium cells is that they are of almost limitless life. They could be charged and recharged a thousand times or more without causing any damage. When using Ni-Cad cells, it's best to discharge them completely before recharging, as they seem to respond better when this is done. Storing them when completely discharged apparently does not damage them. The discharge characteristics of the Nickel-Cadmium cell make them very useful for caving because they maintain almost their full rate of output until they are just about ready to quit. This is not true of alkaline cells in which power loss more closely resembles the steadily decreasing curve of the zinc carbon cell.

Another major advantage of the electric lamp is that it's possible to carry several different types of bulbs with different light outputs (and thus differing current drains). This means that when the batteries get low during a long trip, one can get by with less light but longer battery life if necessary. When resting in a cave, incidentally, most electric cavers switch off their lamps to get a light ride off the carbide caver and thus conserve battery life.

Many cavers have found Nickel-Cadmium cells at surplus sales stores in various parts of the country. These are sometimes a good buy, but be sure that they will charge and recharge before installing them in a permanent cave pack. They are usually about

¾" by 3" by 6" high. Occasionally, check the liquid level to be sure that they are being charged properly. Because the cells are relatively small in size, a good working voltage can be obtained by using four to six connected in series to give multiples of the 1.4 volt basic single-cell voltage. Six, seven and a half, and nine volt are popular sizes, depending on the types of bulbs available. A grouping of four Nickel-Cadmium cells (a nominal six volts) will mount very nicely in a square 50 caliber plastic ammunition case still available in some surplus stores. Other combinations are possible depending on the availability of suitable containers in the local area. A generous sealing of the top of the cells (leaving only the vents clear for the filler tubes) can be made with RTV compound sold in electronics stores. This comes in a tube or can as a pliable rubbery substance about the consistency of toothpaste. When dry, it's still rubbery but offers a great deal of protection both for shock and dampness.

Backup Lights

The backup lights carried by cavers in addition to the main headlamp are most often a common flashlight and a candle with matches in a waterproof container. It's handy to attach a small length of parachute cord or leather thong to the bracket of the flashlight so it can be tied to a belt, wrist, or pack. Some years ago, waterproof flashlights were popular, but there doesn't seem to be any major advantage to these and they're quite expensive.

Most cavers prefer a standard flashlight, such as the Ray-O-Vac Sportsman, in the smaller size using C cells. The C cells have less life than the standard D cells, but the smaller size and weight are convenient in caving since this is not going to be the main source of light, nor would its usage be that expected of a household flashlight. When transporting a flashlight via pack or on your person, reverse one of the batteries so that if the switch is accidentally turned on, the batteries won't drain. Flashlights are very handy underground not only for emergency use, but to spotlight a particular formation or area on a distant wall or as focusing aids when doing cave photography.

Candles are the third source of light almost universally carried by cavers. The most common type are the plumber or household candles available in any store or supermarket. Usually these are cut to an inch or an inch and a half with the wick carefully dug out so that it can be easily lit. Another type of candle is readily made by using candle wax poured into a 35mm film cassette and employing an asbestos wick. To hold the wick in the center while pouring the wax, some use a small dimestore magnet. Matches stored in a waterproof container of the approved Boy Scout or camping type always accompany the caver's candles in his cave pack.

PACKS AND OTHER PERSONAL EQUIPMENT

Almost every caver on his second or third trip ends up carrying some kind of cave pack. These are usually various war surplus packs, but they should have two general characteristics. They should have a good shoulder strap, yet still be easy to carry by hand for times when it's impractical to have a pack dragging alongside in a tight crawlway. Secondly, the pack must close and reopen easily yet securely. Cave mud causes problems with some types of fastenings so this should be kept in mind when looking them over.

The standard type of rucksack or weekend pack is also seen occasionally for carrying caving supplies to the entrance of a cave. However, cavers seldom use these inside a cave because they are too bulky and a single strap over the shoulder is much handier. Rucksacks are also handy for storing caving gear, including helmet, coveralls, and the cave pack itself, prior to the next cave trip so that everything is ready to throw in the car when the call comes. One final thought on cave packs—be sure to prepack and check supplies before setting out on a cave trip. No one likes to sit and wait while two or three cavers sort through the debris of the last trip at the entrance to the next cave.

A very handy addition to the caver's personal equipment is a universal sling made up of 8 to 10 feet of either ⅜ or 7/16 inch climbing rope or tubular white webbing. It could also be a doubled 16-foot length. The advantage to this universal sling is that it can be a waist loop, chest sling, diaper sling, tie-in line, or dunnage line. For a waist or diaper sling it is first tied with a fisherman's knot. For a chest sling, the same loop can be worn around one shoulder and the waist, as shown in Figure 5-5, for quickly tying in when on belay.

Another item of personal gear is knee pads. These are absolutely indispensable in caves where any unusual amount of crawling is to be done. They are available from janitorial supply houses and some large industrial wholesale firms.

A minimum first aid kit should be carried by at least one member on each caving trip. This would include band aids, a

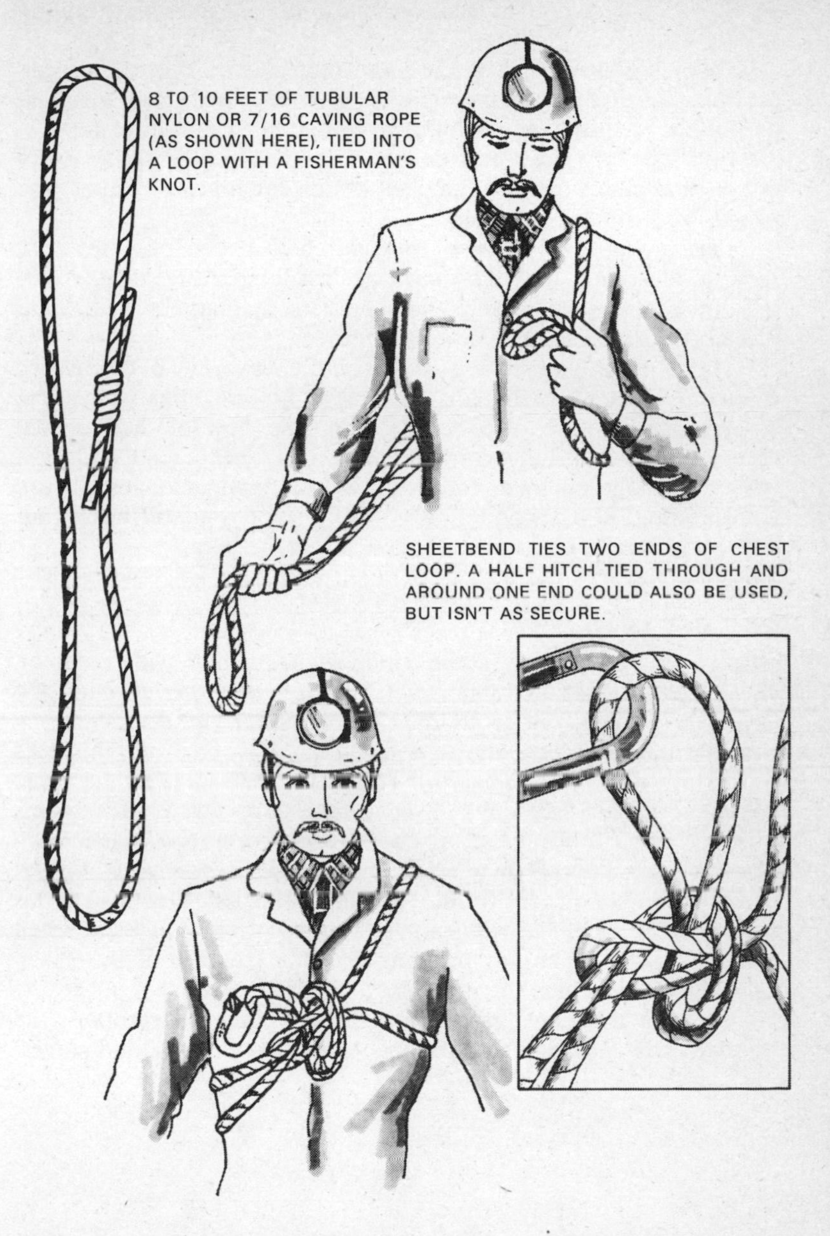

Figure 5-5. Universal Caver's Sling. Worn here as a chest sling for belay line, mechanical ascenders, or ladder-resting position by clipping into a rung. Can also be tied as a diaper sling for rappelling, or be used full length alone or tied to another sling for hauling gear or tying into a safety line when working near a pit or belaying.

couple of butterfly bandages, disinfectant or first aid cream, and aspirin. All of this can be easily carried in a small flat metal or plastic box, tightly secured by a rubber band or a strip of tape.

Finally, every caver carries water in a canteen or plastic water bottle, and many carry a pocket knife, can opener, paper and pencil, and often a compass.

Cavers who wear glasses outside, had better plan to wear them inside caves, too. To guard against loss or breakage, sporting goods stores sell elastic glass holders that attach to the ear pieces and drape loosely around the neck.

There is, of course, the story of the caver who discovered some rare green speleothems in a cave he had already been to several times before. He couldn't imagine how he had missed them. As he was duly contemplating this phenomena and how best to photograph them for posterity, his companion caught up with him and asked wonderingly, "How come you still have your sun glasses on? The sun isn't that bright in here."

Cave Food

Almost all cavers carry some kind of snack food with them on cave trips. Generally, the exact type of food varies with the cavers concerned, but it all has two common characteristics: it provides quick energy, and it is not easily damaged by the rigors of the wet, muddy cave environment. Here are some of the more common snacks seen on a typical cave trip: M & M candies, raisins, Tootsie Rolls, or sugar cubes carried in a stout plastic container; canned meat such as tuna fish, Vienna sausage, deviled ham, or boned chicken; dried fruit like apricots; breakfast foods such as Granola mixed with nuts and rolled in a plastic bag; small canned (snack-packs) of pudding or fruit (beware of the sharp edges inside the cans, though); and beef jerkey carried in plastic.

On long trips involving overnight camping underground, the standard fare is usually backpacker-type foods, utensils, and stoves.

6

Techniques for Horizontal Caving

The basic movements in horizontal caving include walking, crawling, squeezing through tight places, scrambling, and chimneying. Horizontal caves usually descend gradually in a series of steps varying in size from a few feet to fifteen or twenty feet. Or, they may be made up of several distinct levels connected by shallow pits and chimneys. Despite these vertical features, these caves are still considered basically horizontal, even though some scrambling and chimneying are necessary to negotiate them. It's only when one must use rapelling, prusiking, ladders, or belayed climbing that a cave is considered truly vertical.

It should be emphasized again that the purpose of describing techniques in a book like this is to provide a familiarity with the terminology and the basic structure of the technique. Nothing can substitute for actual field experience where every technique is practiced many times until it is completely second nature.

Proper caving technique calls for a special kind of self-discipline combining a state of mind, a degree of physical ability, and a full appreciation of the dangers involved. These three qualities fuse together to make the mechanical rules of technique take on their true meaning when the caver is in the real underground situation exposed to the risks involved.

A word of caution which bears repeating: Never go caving alone! For any cave trip, four is a minimum and if there are beginners, at least two experienced cavers are needed on the trip. A beginner's trip shouldn't last more than two or three hours, at the most, on the first exposure.

Also, keep conservation in mind, since any movement in a cave can cause damage unless care is exercised at all times.

FUNDAMENTALS FOR NEGOTIATING PASSAGES

Cave Walking

Walking in a cave is not really hard when a caver gets the hang of it, but it's surprisingly difficult compared to the more accustomed walk from parking lot to supermarket. It's very much like hiking cross-country away from an established trail, hopping over boulders and outcroppings along the way, climbing small hills, fording streams as they occur, and even walking right down the stream bed, itself, for fairly long distances, this sometimes being the only natural trail available.

As when crossing a stream by stone hopping, a caver always looks before he steps. By lighting the way in front of him, he can plan two or three steps ahead and let his feet follow as in normal walking. Often, he will brace himself with one or both hands on the wall (unless in a fragile area), to keep good balance. If he wants to study the passage or sightsee, he stops, so that his full attention is directed to where he's going while underway.

As in any kind of hiking, pace is extremely important. It's far better to move along slowly and steadily than push ahead in fits and starts. When necessary, take several short rests rather than long ones because the 45-55° temperature usually encountered in caves can easily chill.

If a novice gets tired, he should say so then while he still has some reserve strength. It's not fair to the other cavers on the trip for anyone to get so completely tired out that he becomes a burden.

When walking in stream passages, most experienced wet cavers simply slosh right in, making no attempt to keep their boots and socks dry. This is probably the best way to do it, since the alternative is to try and cling to a wall in some kind of complex

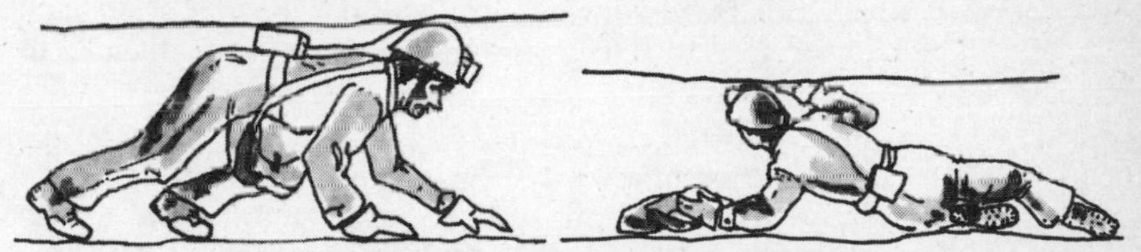

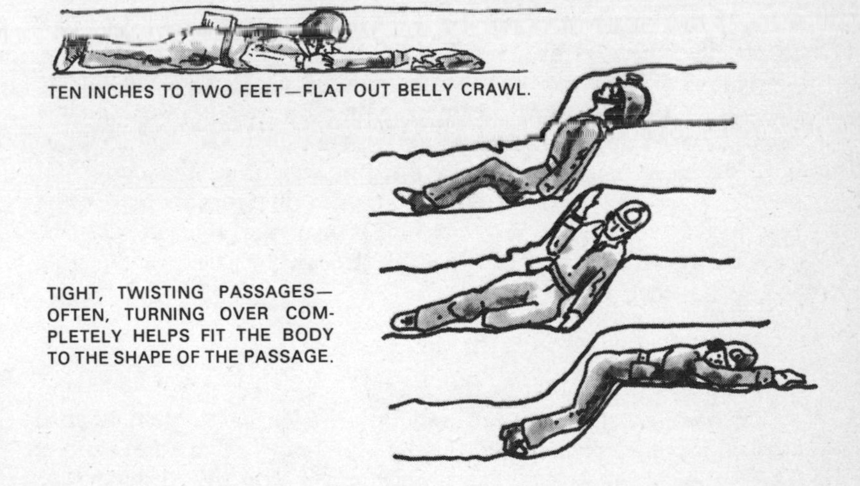

Figure 6-1. Basic Crawling Movements. Experienced cavers study each passage, then more or less pour themselves through the tight spots.

traversing maneuver, which can easily cause a fall or a twisted ankle. However, there is also much to be said for dry feet, so if it's just for a short distance, there's no sense in getting one's feet wet any more than one has to. Obviously, avoiding water falls is also wise since getting wet chills the body and can possibly lead to hypothermia.

Crawling

Crawling isn't the only thing that cavers do when caving. It just seems that way! However, there are those who actually like to crawl, and they think crawling is what caving is all about. In reality, nearly all caves require some crawling, even if it's only for a few feet. This is one extreme. The other is those crawling horrors which call for earthworm-like motions on one's stomach for hundreds of feet with perhaps only one or two chances to sit up.

Fortunately, crawling is very easy to learn since most people seem to do it quite naturally (it seems even cavers were children once). Merely allow atavism to take over and return to the belly-crawling age of 10 months. And this is very proper, too, since the stomach, arms, and feet are the best things to use for crawling. Contrary to what might be a first impression, the knees are *not* the best part of the body to use. In fact, it is wisest not to use the knees at all, because the knee caps are very delicate. A damaged knee cap can be so painful as to cause complete blackout. Furthermore, when crawling on the knees, valuable body heat escapes quite easily. It seems that the human knee was just not designed to carry the full weight of the body. Of course, everyone uses his knees at one time or another when crawling (and even when climbing), but the rule is, use the knees only when necessary! If a lot of crawling is expected, bring along the knee pads.

Getting Stuck

When one first thinks about crawling in a cave, he often is filled with visions of becoming stuck and remaining there forevermore. The truth is, however, that a new caver can get through just about any tight squeeze with practice; the chances of getting stuck are really quite small. Look at it this way—many cavers have been through that particular tight spot in both directions and if they can make it, others can, too. No squeeze is as bad as it looks.

Here are some of the tricks of the trade. When negotiating a tight crawl, the caver first empties his pockets. Then he considers which parts of his body are the widest. Normally, this will be the hips, shoulders, or pelvis. Then he studies the shape of the tightest spot and plans the best way to fit his body through. The experienced cavers kind of pour themselves into the tight squeezes. The important thing is to be as relaxed as possible so that as one eases himself in, he can adjust his body to fit the passage as needed.

If it's really tight, many cavers will take off their helmet and push it ahead of them. The same goes for packs and other items being carried. Often the caver ahead will help the one behind by pulling his gear through for him. One common way to do this is to attach the pack to the foot of the caver next in front (or even to one's own foot) and drag it along. Few crawlways are so tight that a dragging pack will get stuck. Sometimes it helps to narrow the shoulder area by putting one arm ahead and dragging the other arm behind.

Those new to caving would be well advised to find out at an early date just how small a hole they can get through. The rib cage limits vertical movement. The hips and shoulders limit the width. Once the limit is known, no caver should push beyond that limit and so endanger the whole group.

In theory, if the caver can get his head through a tight spot, the rest of his body will go through, too. However, this assumes that the passage is at least a yard wide, is only really tight for a short distance of six to eight inches, and then opens up on either side. Even so, it's encouraging to know that if one's head can fit through, the rest of him, one way or another, can probably make it, too.

Descending Crawlways

A word of caution in crawlways that go down. Never go head first down a steep unknown crawlway. Its far too difficult to back out after going down head first. So unless the crawl is very well known and ends in a flat patform area, don't go in head first. Also, it sometimes happens that the passage bells out suddenly, leaving one with no forward support of any kind.

Begin on the Stomach

In a really tight passage, it's best to start off on the stomach. Push with the toes and pull with the arms while wriggling like a snake with the rest of the body. Even the tightest crawlway normally has ledges and knobs which can be used to push off with the feet and to pull forward with the hands. Whenever possible, try to have the legs do most of the work since they are far stronger than the arms.

The human body bends forward and sideways from the waist or knees fairly easily but it doesn't bend backwards very well at all. Keep this in mind when studying a passage and consider the various rises and twists in the route. Depending on how flat it is, cavers very often turn over halfway through a tight squeeze to take advantage of the natural bending places of the body. Sometimes, there are also footholds that are easily reached when lying in a new position but can't be reached in the original position. See Figure 6-1.

In particularly tight spots, an experienced caver, particularly

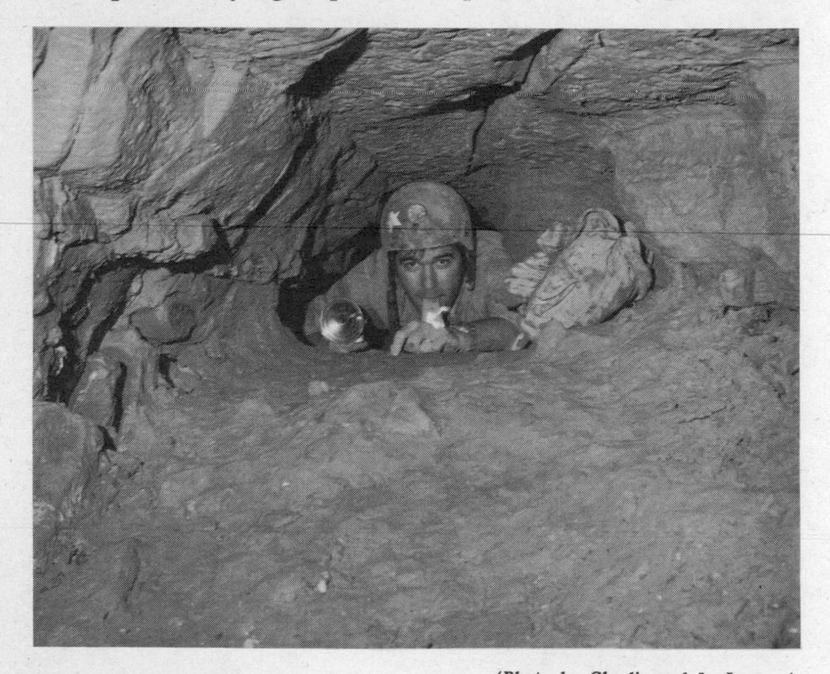

(Photo by Charlie and Jo Larson.)

Emerging from a tight crawlway in Greenbriar Caverns, West Virginia, this caver is carrying a flashlight in his mouth and a carbide lamp in his hand.

the caver that's just been through, should talk the new caver through the squeeze. Encouragement and the knowledge that several people have gone before will almost always get a beginner through.

Freeing a Stuck Caver

In those cases where a caver really gets stuck, there are several relatively easy remedies. First, be sure he really is stuck. Sometimes simply unhooking a pack will free him. If he is stuck, the most obvious remedy is for the other cavers to grab him and pull. This works in five cases out of six. If it doesn't, another successful technique is to run a length of rope through the passageway with a figure eight loop tied into the line so that he can get a good hold on the rope with his hands and the others can pull him out. If this is to no avail, he or a caver on the other side can place the loop around one of his feet. Then as soon as the loop is attached, he raises his knee and the others take up the slack and continue holding the line under tension. This allows him to use the line as a foothold and push hard with his foot until he moves forward two or three feet. He then raises his knee again, the others take up the slack, and he repeats the process. Surprisingly enough, this will free the worst stuck caver.

In any crawlway known to be tight, a hand line with or without loops can often be carried through by the first man for the others to use for hand or foot holds.

Throughout a crisis of this sort, it's very important that the other members of the group do everything they can to keep up the stuck caver's spirits. This can't be emphasized too strongly, since depression often leads to exposure and a feeling of malaise that may help a caver convince himself he's stuck forever. If it does, he may give up and one will have a real problem on his hands. So, keep up a constant stream of reassurances, jokes about how many times various ones in the group have been stuck, and other light talk that will keep his mind off his present predicament.

Sometimes it's a good idea to have one of the thinner members of the party crawl back in and stay close to the stuck member to give him reassurance or pass him some food or drink. The main purpose is to have him relax so that his muscles are loose so he can slide through easily.

A not uncommon experience among cavers is the one who seems so irrevocably stuck that pulling on his arms or legs causes no movement at all. Then suddenly, he relaxes (or lets out his breath) and out he flies, with enough force to knock all his rescuers over. It often happens that the stuck caver condition ends in this type of jocularity. Hopefully, the victim gains in confidence what he may have lost in weight.

Types of Tight Passages

Although this has mostly dealt with horizontal crawling, tight squeezes in caves can be found in vertical slots or in sloping fissures also. Either of these often have tight spots that require the same basic techniques as tight crawls.

Horizontal passages can be classified according to their height as shown in Figure 6-1. Passages from four to six feet high can usually be walked with some stooping. It may help to place the hands on the knees where possible to rest and maintain balance.

In passages two to four feet high, several methods can be used. Many cavers alternate them to avoid tiring. The most obvious one is to progress on all fours, avoiding the use of the knees whenever possible. In a short passage, some cavers prefer to go very fast on all fours just to get through, but this is strenuous and difficult to do for any distance. Another possibility is to use a crouching walk on the haunches when the arms get tired of movement on all fours. This is also good position for resting occasionally. A good way to avoid the use of the knees is to travel on the elbows or forearms and the toes of the boots or ankles if the feet are turned on the side. Sometimes lying on the side is preferable in passages of this height, shuffling the feet and pulling forward with the arms. Another alternative is to tuck one leg beneath and squirm forward in a kind of inching movement using the hands for balance.

In passages from ten inches to two feet high, the best method undoubtedly is the flat-out belly crawl. This is very much like the technique used to crawl under a bed to retrieve a slipper. The elbows are flat out to the side of the body, the pack is pushed in front, and in many cases the helmet has to be removed and pushed ahead also. However, movement in this manner is surprisingly easy if the floor is not too rough. As long as the crawl is not more than thirty feet, this kind of motion is not too difficult to carry out except when the ceiling gets down to eight or ten inches and it really gets tight.

The problem in caves is that passages don't have just one height. They vary from place to place and so therefore do the techniques that a caver must use to traverse them. It is also a fact that any two cavers may use a slightly different technique in traversing the same crawlway because what suits one because of the length of his legs or arms or trunk may not suit another because of his different dimensions. It's necessary for everyone to find out for himself just how small and what shapes of passage he can best maneuver, remembering to keep as loose as possible and never forgetting that he can always get through, one way or another.

Scrambling

Scrambling is a technique used when a passage is too steep to walk up, yet not steep enough to require classic rock climbing skills. Scrambling combines several types of movements, including walking, climbing, sliding, and jamming. Generally speaking, any downward or upward cave passage of less than 45 or 50 degrees can be negotiated by scrambling. If it's steeper than that, chances are it will have to be climbed using more or less standard rock-climbing techniques. But a surprising number of slopes that at first glance seem very steep can be easily scrambled using a combination of balance, three point suspension from rock climbing, and the seat of one's pants.

Another word about conservation at this point. Any movement in a cave can cause damage, but this is especially true of scrambling. Avoid scrambling techniques in formation areas where delicate speleothems could be destroyed by a careless movement. Use an alternate technique like roped climbing or seek another route.

The new caver should not be surprised if he is frightened when negotiating his first 10 or 15 foot slope in a cave. Remember the maxim of mountaineering. A rock climber without fear is dangerous. The same thing goes for cavers. Any caver who is not afraid, and therefore cautious, is dangerous. The foolhardy endanger not only themselves but everyone else on the trip.

When scrambling down, gravity and friction are on the caver's side, so it's usually easier than going up. The trick is to move only one limb at a time leaving the rest of the body to hold fast in a secure position. This is the so-called three point climbing technique used by rock climbers and it's quite easy to adapt to scrambling down a 45° angle. Use the entire body, arms, legs, feet, back, shoulders, seat, and even one's helmeted head once in a while for balance. Then, stretch an arm or a leg down to the next handhold or flat area keeping the other three in close contact until the new hold is secured.

As with all of these techniques used underground, it's best to practice out of doors to get a feeling for the basic technique. Almost any area has some large boulders along the shore of a lake or in nearby hills. Gradually descending blocks are very similar to the type of pitch found in caves and make excellent practice for learning how to scramble in a cave.

When scrambling, it is not necessary to maintain a careful upright position with one's head directly over his feet as used on steeper pitches in rock climbing. In fact, on slippery slopes it helps if one almost lies down using his legs, seat, back, and shoulders to give him more friction.

In scrambling, as in climbing, the arms are used primarily for balance. This means that the legs support the weight, not the arms and hands, although for a short time of a few seconds up to a minute maximum, one can use an arm or hand if necessary. As in any kind of climbing, never lunge or jump. Speed is not the object. Plan the route of movement ahead, checking each move while proceeding and the next level will be reached in no time.

Climbing back up a slope is more strenuous, but is really not much more difficult than climbing down assuming the caver is not too tired. One caution—if the slopes are *up* on the way in the cave rather than down. The rule in rock climbing is that one must be able to climb back down anything that he climbs up in case it runs into a dead end. The reason is that in climbing, going up is often much easier than climbing down, because it's easier to see where you're going and find holds, although this is not as true for scrambles as for free climbs. Nonetheless, it's a precaution to keep in mind when the cave tends to go mostly upwards on the way in.

Handlines and Belays

Many times in even the easiest cave it's a smart idea to rig a handline for the use of new cavers on those 10 to 20 foot scramble pitches. Any standard caving rope will do very nicely for the purpose as will a length of one inch tubular nylon. On extremely steep or muddy scrambles, tie a series of figure eight loops in the line to provide extra hand and foot holds. While the overuse of handlines in caving is not to be encouraged, since any rigging takes time, there is an essential difference in philosophy between caving and rock climbing. That is, climbing (and scrambling) in caving is not an end in itself, as it is in rock climbing. Rather the end of the cave is the end that a caver seeks. For this reason hand lines, ladders, and other types of mechanical aids are much more commonly found in caving than in classic rock climbing.

When caving with very new or less experienced cavers, never refuse to rig a handline if someone asks for it. Safety is more important than bravado. It is the senior caver's responsibility both to the inexperienced caver and to himself in case he becomes injured to observe all safety precautions until everyone understands his own capabilities and limitations.

Chimneying

A chimney in a cave is just about what would be expected from the name. It is a narrow vertical passage, sometimes tubular, sometimes a fissure, sometimes just where two walls come together. Chimneying is the technique used to either move down or up between such walls (the horizontal movement in a traverse is also quite similar). In Britain, chimneying is sometimes referred to as "back and footing." Actually the British term is quite descriptive because it is the back and the feet that do the work, with the arms, as usual, being used for balance. Chimneys range in size from about eighteen inches to forty or more inches in width. Chimneying tends to be much more common in caving because cave chimneys are more common. Many cavers routinely shimmy up and down thirty or forty feet without a thought in chimneys that might give pause to a mountain climber.

The exact technique used in each chimney (see Figure 6-2) depends a lot on how wide the distance is between the walls. Starting with descending the really narrow chimney first (about 14 to 24 inches), these are really fairly easy to negotiate because gravity does the work and friction keeps one from falling. However, they are quite a bit harder to climb up, because the legs can't do their work as well.

When a chimney narrows to about 24 inches, put one foot under the body on the back wall and the other foot on the facing wall. This tends to require more downward pressure with the hands but it's still easy to do.

When climbing back up a really narrow chimney, an aided ascent can be used with rope loops tied into a line for footholds. The technique is to raise the foot and have the cavers above take up the slack in the line. Then the chimneyer stands in the loop

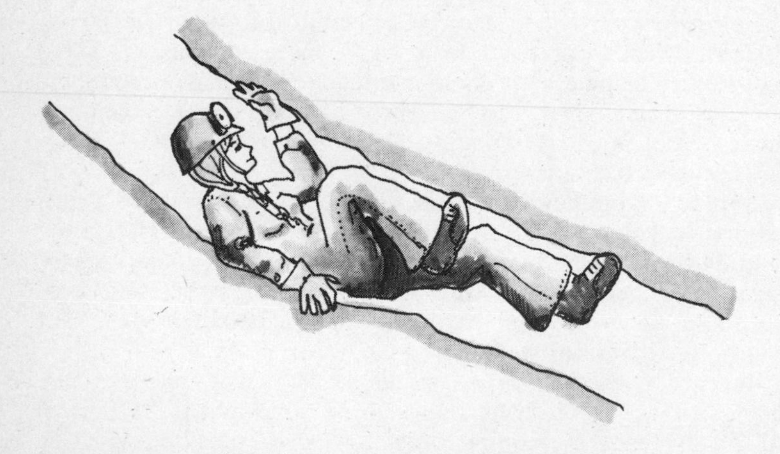

Figure 6-2. Chimneying. Proper technique depends on chimney width. Basically, the feet and back press on opposite walls and the caver ascends or descends somewhat like a spider moving on a wall. Chimneying is useful for vertical and semi-vertical pits and for inclined slides that are too slippery to offer good hand and foot holds.

raising himself two or three feet to a fully standing position. Then he raises his foot again while bracing himself in the chimney (which isn't hard if it's really tight) and repeats the cycle. In this way he can move quickly up a chimney too tight for regular chimneying techniques.

In wider chimneys, the basic technique is to put both feet on one wall and the back on the other wall. Generally, it's best to put the back against the smooth wall since this is easier on the spine in case there are knobs or rough spots.

Basically, what is required is to brace firmly with the back against the wall and move the feet one at a time, while shuffling up or down with the back. For better control, try to keep the knees bent a little, if possible, like a skier or dancer, rather than having them straight out. Also, keep the knees at about waist level but not any higher, otherwise the seat may slip down.

The smooth chimneying motion of an experienced caver looks a little bit like a spider with one limb moving after the other. When a rest is needed, simply put the feet about level with the seat or waist. A caver can stay in this position safely and comfortably as much as three or four minutes without any strain.

Most cavers find that chimneying, while strenuous, is really quite easy with a little experience. It's even possible to practice chimneying at home by straddling a wide doorway or two trees and inching one's way up. This is actually a little bit harder than in a cave because the average doorway is only six or eight inches thick and offers less surface for the back and feet than cave walls.

A modified chimneying technique can also be used to slide down a smooth floored slope when the roof is close to the floor and irregular enough to offer some kind of handholds or footholds even though the floor itself may be of sand or mud and allow no purchase. By simply lying down with the back on the floor and feet on the ceiling, a caver can negotiate a difficult slope impossible to climb any other way. A word of caution again here—never descend head first in a slope like this because it may be impossible to get back out.

Traversing

Traversing is a sideways movement somewhat similar to chimneying, if the walls are close together, or similar to free climbing, if they aren't. In either case, the movement is horizontal rather than vertical. In caving, traversing is often done while straddling a narrow passage, but it can also involve stepping sideways along

Figure 6-3. Traversing along a ledge on belay.

a narrow ledge, using hand holds on the wall as in rock climbing. Often times in a cave, traverses are used to avoid climbing down to floor level and then back up again when it's clear that the passage simply dips down for a short period of time. It's also quite common to use a traversing technique along the walls of a passage to aviod water down below, assuming that this can be done relatively safely and that eventually the route leaves the stream. If not, as indicated earlier, it is just as well to slosh through the water rather than try to hang on to the wall merely to keep the feet dry. Sometimes a straddle technique is used in a cave because the passage is too narrow down below and it would require too much physical energy to try to slither through the thin passage, it being simpler to make a traverse at a higher level.

When moving on a ledge, the best technique is to move the feet in a shuffling motion. Don't try to cross one leg over the other if the ledge is especially narrow and if facing the wall. Shuffling the feet tends to keep the balance, while crossing the legs causes tripping.

Traverses involving very narrow ledges or only a series of small foot holds, should always be rigged with a handline and each caver should clip-in to it from a chest loop with a carabiner. On a particularly dangerous traverse, the first person to cross will be rigging the handline so he should be belayed with a safety line.

In most cases, a traverse will have been negotiated before by others in the party so that a new caver can get advice on where to place his hands and feet for the best movement. Traversing is similar in some respects to free climbing, but it frequently happens that the walls are close enough together so that the back can be placed against one wall or the hands on the other to give additional balance. The trick in moving in a situation like this is to use the hands, back, legs, feet, and seat in any way necessary to provide good balance. Good traversing, like good climbing and good chimneying, requires a caver to be somewhat of a contortionist, pouring his body against the various surfaces and holds that are available.

Carrying a pack or other gear is very difficult when traversing (or chimneying), so it is best to work out a method of passing gear along to the next man or lowering it. In a cave under intensive study where traffic along a certain route may be heavy, it's common to rig a more or less permanent hand line using

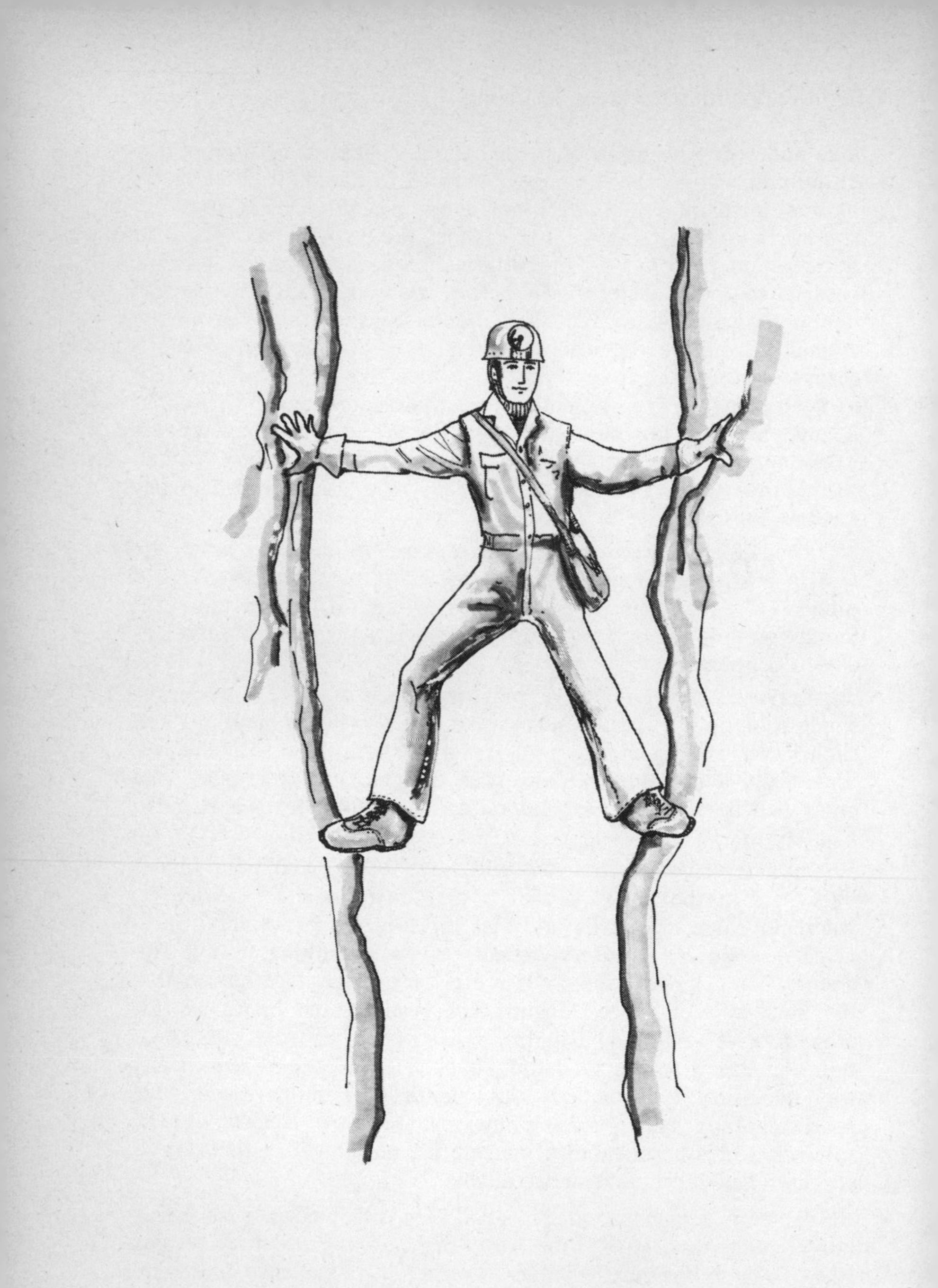

Figure 6-4. Straddling a fissure, another type of horizontal or traverse movement. This is also called canyon hopping.

bolts and bolt hangers so that the trip is speeded up and dangers are minimized.

IMPORTANT KNOTS

Knot tying is one skill a new caver should learn at home rather than in the field. In fact, it is his responsibility to know the basic knots like an expert before he sets foot in a cave. Nothing is more discouraging than to get ready for a drop and then have to spend several minutes showing each novice how to tie in. When learning knots, a good method is to use small diameter cord of two different colors so it's easier to follow the formation of the knot. Then, transfer this technique to 7/16 or ½ inch nylon and practice until it's second nature.

Types of Knots

In caving, knots are used for four main purposes. First, to tie the caver in at the end of the line; second, to attach two ropes together; third, to form a loop to use as a foot or hand hold, or to clip equipment into; and fourth, to provide a means of safely climbing a fixed rope in a technique called prusiking.

The minimum essential knots for caving are the bowline, the overhand knot, the fisherman's knot, the figure-of-eight loop, and the prusik knot. A caver must be able to tie these knots in the dark. Rock climbers often insist that their novices learn to tie knots behind their back with both hands. This isn't a bad idea for cavers and they should practice this, too.

Since all ropes used in caving are of artificial fibers, the inflexible rule is: secure any knot tied in any man-made rope with two half hitches or two overhand knots.

A note of warning about the square knot—a square knot is used for tying packages, but it is dangerous as a caving knot. The problem with the square knot is that it turns into a slip knot (a granny knot) when it is tied wrong and it is extremely easy to tie it wrong! It would be best if cavers forgot about the square knot completely.

One other general point about tying any knot. Be sure to form the knot carefully as it is tied and be sure that it retains its shape and form as it is tightened. If one is not careful about this, the knot will probably loosen under tension.

The basic parts of a rope for knot tying purposes are the end,

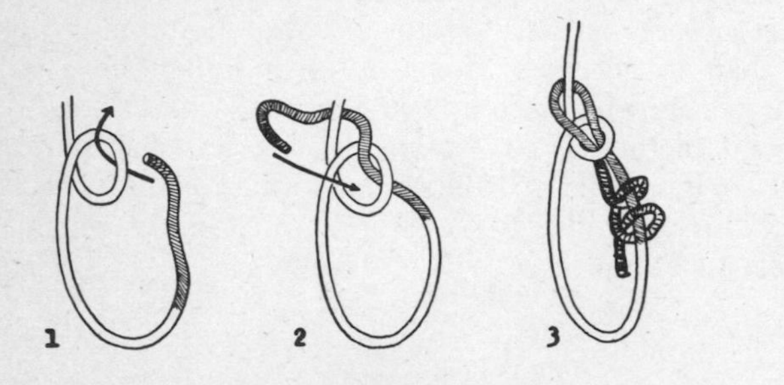

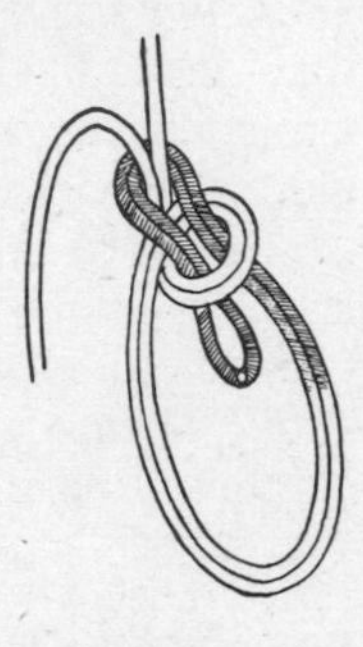

BASIC BOWLINE

Always secure *any* knot in a man-made fibre with two half hitches or preferably two overhand knots.

DOUBLE BOWLINE

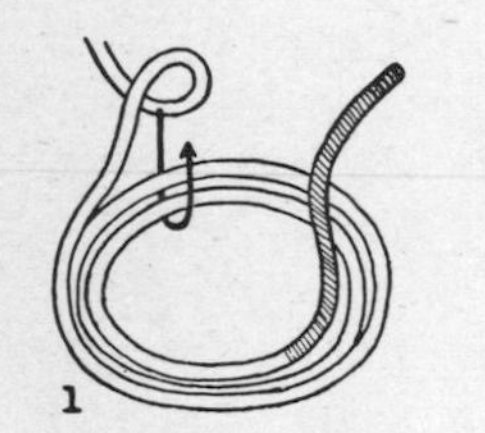

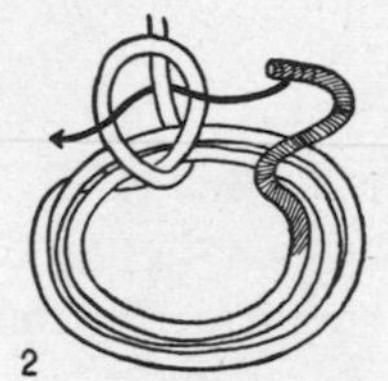

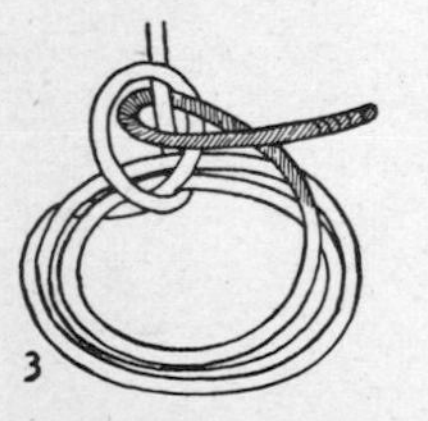

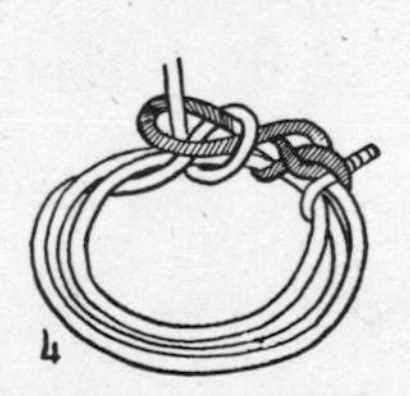

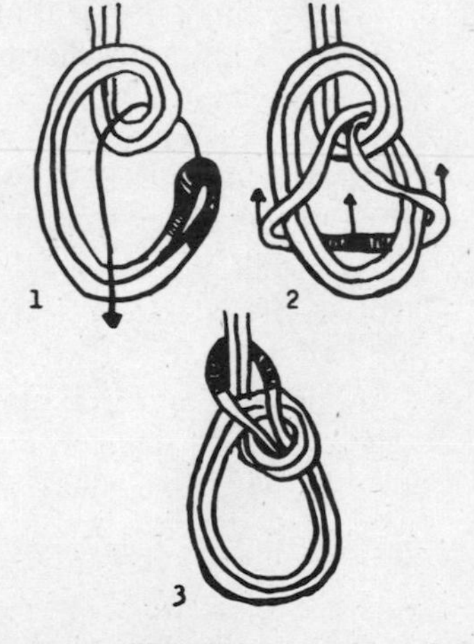

BOWLINE ON A COIL

For tying into a belay line. Tie around chest, not waist.

BOWLINE ON A BIGHT

Figure 6-5. The Bowline Knot in its various forms. This is the most important single knot for any caver or climber to learn.

the standing part (where the knot *isn't* being tied), and the loop, which is formed by placing the end over the standing part.

The Bowline

If a caver could only learn one knot, the bowline would be it. A bowline can do three of the four things that caving knots must do. It can provide a non-slipping loop at the end to tie the caver into the line or to attach a line to an anchor; it can tie two ropes together by simply running one loop through the other; and it can provide a loop in the middle of a rope for holds or equipment.

A bowline is relatively easy to tie, but there is only one correct way to tie it. Follow the diagram as shown carefully, so that the end of the rope finishes off inside the loop. It is possible to tie it so that the end of the rope ends up outside, thus weakening the knot significantly.

A double bowline can be tied either in the middle of the line as a non-slipping loop, or at the end, if two loops are desired. A bowline on a bight is constructed similarly, as shown.

The best way to tie a bowline is to first lay a loop *over* the standing part of the line. Then, thread the end of the line up through the loop, around the back of the standing part of the line and back down through the loop again, ending up in the center of the main loop. Tighten it carefully and always secure it with two overhand knots. When tying a bowline around an anchor, the knot tier stands with his back to the anchor and the standing part in front of him. Run the line around the anchor behind and tie as above.

The bowline on a coil is preferred around the chest for tying into a belay line, because the extra coils reduce the shock to the body in case of a fall. Take about 10 or 12 feet of line and while holding the standing part in the left hand, wind the end around the body four or five times in a counterclockwise motion from right to left with the right hand. Then while still holding the standing part of the line in the left hand, form a loop *over* the standing part of the line with the right hand and snake this loop up underneath the coils that circles the body, being careful not to lose the twist in doing this. Center this loop tightly against the standing part, then run the end of the line through this loop around the standing part and back through the loop once again. Before inserting the end in the loop be sure to slide the coils counterclockwise around on the chest to tighten them. If not tight enough, the whole thing may pull right on up over the arms and shoulders in a fall. Finish off with two overhand knots around all the coils.

Overhand Knot

This is simply a loop over the standing part of the rope through which the end is placed. Two of these are tied around any synthetic fiber rope to secure the main knot. By doubling the rope, an overhand loop can be made in the same way, but it is not recommended because it is much harder to untie than the figure-of-eight loop.

Prusik Knot

This is quite an unusual knot, developed originally for glacier rescue work. It is tied around a fixed line and will hold tightly without slipping even when downward pressure of up to several hundred pounds is put on it. But it is also easily loosened and slid up on the rope when the pressure is removed. It is used for prusiking, a means of climbing a rope safely with foot and chest slings that are attached to the rope by means of the prusik knot. It can be tied as shown in its four-coil form and will hold well on most ropes, including muddy ones. It can also be tied with one or more turns in a six-coil version, if the rope is especially slippery.

Besides its use in prusiking, it is a good knot for tying into a line as a self-belay or safety when working in a steep or hazardous area. It also finds use in hauling gear and in rescue work.

Fisherman's Knot

The fisherman's knot is very simple to tie. Equally important, it is relatively easy to untie if it was tied properly. Basically, all it is, is an overhand knot tied over one line and then a second overhand similarly tied in the second line. Both knots are then snugged together tightly. However, as with all knots in synthetic rope, it must be secured by either another overhand or two half hitches to be completely safe.

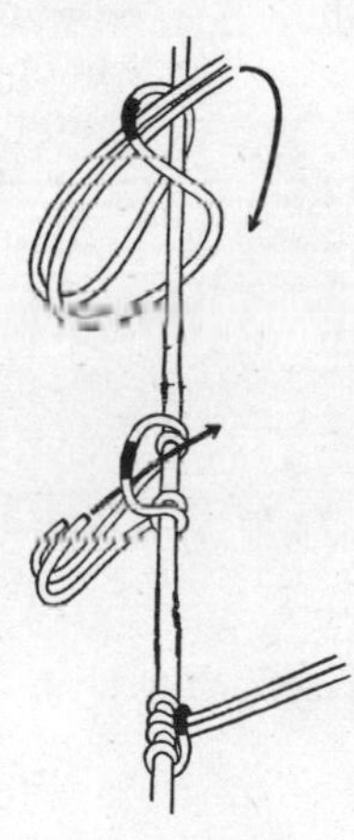

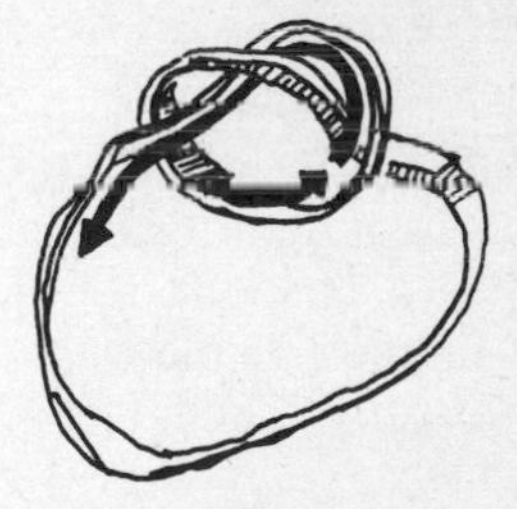

Figure 6-6. Other Basic Caving Knots.

Figure-of-eight Loop

The final basic knot for a caver is the figure-of-eight loop. This is preferred to a simple overhand loop because it is far easier to untie. It simply requires one more motion back over the top of the rope to tie, as shown.

Sheet Bend

The sheet bend, sometimes called the weaver's knot, is usually specified as a knot for tying together two ropes of unequal size. Actually it's quite good for ropes of the same size, too, and one of its main advantages is it is quite easy to untie. For this reason it's often used for tying slings of tubular or flat nylon. Like all knots it should be secured with two half hitches or two overhand knots on both ends.

Half Hitch and Two Half Hitches

The half hitch is a loop around the standing line, two half hitches being two of the same. These knots are not safe enough to use for joining two ropes together but they are commonly used to secure the ends of knots in synthetic rope. However, two overhand knots are preferable, being less likely to loosen up, and they are recommended over two half hitches for caving use.

Water Knot

The water knot, also known as the overhand bend, is easy to tie and is good for tying webbing of the same or differing size. Simply tie a straight overhand knot and then thread the other rope through in the opposite direction, again securing both ends with two overhand knots.

Weakening Effect of Knots

One of the things to remember about any knot is that because of the bends and short radiuses involved, a knot will weaken the rope by a factor which runs as much as one third to one half of its strength. The accompanying table shows how much strength is left in a rope when tied with certain knots. The key thing to remember is that while any knot weakens the rope, one knot may do so less than another. For example, the sheet bend is recommended as an alternate knot to the fisherman's knot because the efficiency of the sheet bend is 65% as opposed to 59% for the fisherman's knot. Other percentages are shown in the table.

The weakening effect reduces the strength of a piece of 7/16 inch climbing rope from 5000 lbs. to about 3000 lbs. This is still a considerable margin of safety, but it is one of the reasons why manila ropes which only have a strength of ⅕ to ¼ of nylon are not recommended for caving.

Weakening Effect of Knots
(% of original strength remaining, ¼" nylon)

Knot	%
Bowline	65%
Fisherman's Knot	59%
Figure-of-eight Loop	50%
Sheet Bend	65%

BEGINNER'S QUALIFICATION PROGRAM

After a tyro has had some training on several grotto or club trips, his basic skills can be checked out using the following suggested program:

1. Tie the four basic knots blindfolded or in complete darkness, the bowline (in one or more of its forms), fisherman's knot, figure-of-eight loop, and prusik knot.

2. Crawl through a tight passage, 12 inches or less in height, two feet or less in width, and ten feet or more in length.

3. Crawl through a passage of similar dimensions with at least one "S" curve that requires turning over or crawling on his side, with only one shoulder ahead of him, and pushing hard-hat and gear ahead of him.

4. Slide downward (in basically a chimneying movement) through a semi-vertical passage (less than 45°), or fissure, that is not more than two feet in diameter and at least 10 feet in total depth. (This assumes the landing below has previously been explored and is known to be a safe stopping place.)

5. Similarly, crawl, chimney or climb back up this or a similar passage.

6. Slide down a passage or fissure of similar dimensions as above, but one that is tilted sideways, so that he must slide, not straight down, but at an angle.

7. Similarly, climb (or clamber) back through this or a similar slanting fissure.

8. Chimney down, while on belay, a vertical or semi-vertical pit (more than 45°), that is wider than two feet on the average, and at least 15 feet deep.

9. Similarly, chimney up the same pit.

10. Make a horizontal traverse, while on belay, using holds as well as counter pressure, if the walls are close together.

11. If using a carbide lamp, extinguish it in total darkness and immediately find and light his secondary source of illumination. Then, determine the cause of the following troubles and correct them. (The instructor will purposely introduce some of these troubles in the lamp or another lamp.) This test assumes the new caver to be carrying extra water, carbide, and a spare parts kit with him *at all times* in the cave, plus the normal two other sources of light.

Lamp won't light—No water or carbide, clogged tip, wet felt, bad gasket, loose bottom. The novice under test must refill the lamp with water and carbide, and put used carbide in a suitable container for removal from the cave.

Lamp burns irregularly—Same causes as above, but he must dismantle lamp and actually replace felt and tip.

Flame around gasket—Bad gasket or thread seat, loose bottom.

Flame around tip—Bad or loose tip, bad tip seat.

Water spurts from tip, or bubbles out of hole in water filler cap, or flame is much too long—Too much water; decrease flow and wait before lighting. Felt may also have been soaked and need replacement.

12. If an electric lamp is used, similarly, turn it out, find and light secondary source for illumination. Then determine the cause and repair the following troubles. It is assumed examinee has spare bulb and batteries with him at all times, as well as two other sources of light.

No light—Bad bulb, dead batteries, loose connection. Take lamp apart and replace bulb and batteries.

Irregular light—symptoms as above.

Dim light—Check batteries for corroded or loose connections.

Cable catches on obstructions—Disconnect cable and reroute from battery to lamp. Also demonstrate ability to disconnect the cable quickly when it snags in a tight passageway, chimney, or fissure.

13. To check for conservation skills and awareness, here are two additional tests to be conducted above ground in full caving gear:

a) Crawl through a man-made obstacle course of chairs and other furniture covered with loose objects that will easily fall off, without hitting anything.

b) Walk across a floor densely strewn with ping pong balls without touching any of them.

When one can demonstrate these skills, he has joined the ranks of the caving fraternity for the majority of caves. Now he can safely try to master the needed skills for vertical movement to qualify for advanced caving in pits and vertical caves.

7

Equipment for Advanced Caving

In caving, advanced equipment means primarily the specialized gear used for vertical caving. This includes ropes, slings, climbing hardware, and cable ladders. Much of this equipment has been borrowed from rock climbing, but there have also been some significant developments by cavers, particularly in the area of prusiking devices and low elasticity ropes.

CAVING ROPES — GENERAL

Ropes are used in caves for two main purposes. The first is for protection, the second is for direct climbing. Protection is provided by a technique called dynamic belaying. Climbing is done directly on the rope, using techniques adapted from mountaineering but now probably more highly developed by cavers, called rappelling and prusiking.

Caving and rock climbing ropes have different characteristics. Generally, climbing ropes need much greater stretch to protect a falling lead climber. Long falls are not all that common in caving, and in fact, a highly elastic rope can be a distinct disadvantage in rappelling, prusiking, and hauling heavy equipment. A caving rope is probably the most expensive single item a caver will purchase. For this reason ropes are quite often purchased by two or more cavers or by the local club from dues or a special rope purchasing fund.

In recent years, there has been a much greater interest in safety, both by cavers and rock climbers, resulting in concern over the type of rope and equipment used. Because a caver depends on a rope for his life, the safety of the rope becomes of paramount importance.

Safety, when discussing a rope, translates into total breaking strength. This strength is modified in the cave environment because of the water, mud, and abrasion that ropes are subjected to. Also, one mustn't forget the weakening effect caused by a knot or the sharp bend where a rope passes over a carabiner. These are estimated by the Plymouth Cordage Company (manufacturers of Goldline rope) to reduce the strength of a rope to about 55 or 65% of its original breaking strength. From a practical standpoint, it's wise to consider this as about one half or 50%. This means that a rope testing at about 6000 lbs. breaking strength, as most modern caving ropes do, really only has a working strength of 3000 lbs. But even this must be diminished by a safety factor. Generally, testing organizations rating ropes consider that a safety factor of at least twelve is necessary for ropes handling human beings because of the shock on the rope when catching a fall. (A safety factor of only five is specified for hauling ropes.) So, with the factor of twelve multiplied by an average caver, weight 200 lbs., this comes to 2500 lbs. in round numbers, or near the working strength of a rope. This suddenly makes 6000 lbs. not a nice comfortable margin, but close to a bare minimum.

TYPES OF ROPES

Manila Rope

Manila rope is unsafe for caving. Don't use it! Manila has been universally replaced by the synthetic fiber ropes, overwhelmingly nylon. The only reason that manila was ever used at all, is that it is considerably cheaper than nylon. However, manila has several damaging defects. It is only about one half or less as strong as nylon and it loses its strength even more as its natural chemicals

Figure 7-1. Caving Ropes. Breaking Strengths are 6200 lbs for Goldline (above), and 7000 lbs for Blue Water II (below). Goldline is a mountain laid nylon climbing rope, widely used in caving. Blue Water II is a special low stretch rope made specifically for caving. European alpine ropes, such as Edelrid and Mammut, use the same core and sheath construction as Blue Water.

dry out. It gets very heavy and stiff when wet. Furthermore, it is very prone to rotting and mildew which do not affect synthetic ropes at all. This means it can't be stored wet, but must be carefully air dried. Common chemicals such as battery acid, gasoline, or motor oil also cause deterioration. For these many reasons, it must be stated unequivocably that manila rope has no place in caving for employment as a belay, climbing, or hauling rope.

Some cavers continue to use manila 5/16″ or ⅜″ diameter for prusik slings. However, synthetic fibers have almost replaced manila entirely, even for this limited purpose.

Nylon Rope

Nylon ropes, as used in caving and mountaineering, were originally developed for the U.S. Army Mountaineering Corps during

World War II. Nylon ropes gain their tremendous strength because each strand is made up of a series of continuous filaments running the entire length of the rope. Manila rope, on the other hand, consists of many short strands twisted together.

Nylon rope is constructed in two ways for climbing and caving purposes. The older method is called "laid rope" in which three strands are twisted tightly together, usually with the strands spiralling upwards to the right when held up vertically. The most outstanding example of this type of rope is Goldline from the Plymouth Cordage Company. It is a special mountain laid rope specifically designed for climbing use. Plymouth Cordage also manufactures a non-mountaineering line which is white and called simply Nylon. Both have almost identical characteristics, but Goldline is specifically designed for climbing. Sometimes the white nylon is available at a bargain price so it pays to watch for it.

Kernmantel Ropes—Edelrid and Mammut

The second main type of construction for nylon ropes is called Kernmantel or "core and sheath," which originated in Europe. In the U.S., Bluewater II, a rope designed specifically for caving, uses this same type of construction. The most common European brands seen in the United States are Edelrid and Mammut.

Core and sheath construction uses an inside core of continuous filament nylon which is braided together usually in either loose strands or in a three-strand laid construction. Over the entire central core, is a protective sheath made of braided nylon which holds the core together and protects it against abrasion or chemical and ultraviolet deterioration. One of the main benefits of the core and sheath construction is that the core strands are laid in such a way that there is virtually no twist or kinking of the line. This no-twist quality is extremely important in caving rope where long rappels and prusiks are increasingly common. It is also somewhat easier to coil or braid core and sheath rope because they avoid the tendency to kink.

Generally the core and sheath ropes have greater tensile strength than Goldline, testing out at 7000 lbs. for 7/16 inch Blue Water II as against 6200 for 7/16 inch Goldline.

Certain special high stretch types of core and sheath rope are available for rock climbing. Called Dynamic, these stretch nearly 80% more than their original length, a great aid when catching the fall of the climber on long pitches. However, this kind of fall is seldom encountered in caving. For that reason, Dynamic ropes are not recommended, especially since they are generally about 20% weaker than their standard counterparts.

Recommended Size and Length

For caving use, 7/16 inch (11 mm) is the size recommended. While ropes like Goldline come in a 3/8 inch size that weighs somewhat less (about 2 lbs. per 100 feet), it is also significantly weaker. It tests at only 3700 lbs. rather than 6200. Except for extremely long lengths where weight is a factor or if no dynamic belaying is required, it is recommended that the 7/16" be standardized on since it is simpler to have only one universal size that can be used for all purposes. The same is true in sling material where it is recommended that the novice standardize on one inch tubular nylon webbing for all of his sling needs.

As far as length goes, climbing ropes are usually sold in lengths of 120, 150, 165, or 300 feet. For general caving use, any of these lengths is fine. In fact, if the caves usually encountered have only 20 to 30 foot pitches, a 60 foot length of rope is often more than adequate and a lot easier to carry.

Goldline

Goldline, manufactured by Plymouth Cordage Company, was far and away the rope most often seen in caving, and is undoubtedly still used by more cavers than any other rope. The reason for this is that it is pretty much a universal caving rope. It can be used for belaying, prusiking, rappelling, hauling, and just about any other job that a rope will ever do in a cave.

Although Goldline is rather stiff and somewhat more difficult to coil or braid, after some use, the surface becomes soft and quite pleasant to the touch. This fuzzy surface is actually caused by abrasion and it helps protect the surface against any further abrasion. Be careful when tying knots in Goldline or any other laid rope. The stiffness can make it hard to form the knot properly and keep its shape as it is tightened.

Blue Water II

Blue Water is the first rope developed specifically for caving by cavers. In Alabama and Georgia, in particular, where vertical

CAVING ROPES AND SLINGS

Rope	Diameter or Size (in.)	Circumference (in.)	Construction	Lbs/120′	Breaking Strength (B/S in lbs.) Dry	Breaking Strength (B/S in lbs.) Wet	Strength over Carabiner (% of B/S)	Maximum Elongation at B/S†	Price per Foot	Remarks
Goldline	7/16	1¼	Mtn. Laid (Three strand twisted)	6	6200	5300	NA	50	$ .18	General caving and climbing use, incl. dynamic belaying.
Blue Water II	7/16	1¼	Core and Sheath	6	7000	7000	78%	50	.15	Rapelling and prusiking in caving and climbing.
Kernmantel (Edelrid)	11mm	1¼	Core and Sheath	5	5500*		NA	55	.23	General caving and climbing, incl. dynamic belaying.
One inch Nylon Webbing	1 x ⅛	—	Tubular	3	4000		—		.09	Rappel, waist, chest, seat, and tie-off slings. All purpose sling material.
Two inch Nylon Webbing (Seat Belt)	2 x 1/16	—	Flat	5¼	6000		—		.15	Rappel and various body slings.
Polypropylene	¼ 5/16 ⅜	¾ 1 1+	Laid or Braided	—	1400 (¼)		—		.12	Prusik slings.
Manila	5/16	— 1	Laid	—	1000		—		.04	Prusik slings.
Blue Water II	5/16	1	Core and Sheath	—	3500	3500	78%	—	.12	Prusik slings.

*Estimated. European ropes are not rated in total breaking strength, but in resistance to breakage when holding a fall.
NA: Not Available.
†(% of original length)

caving has developed to such a high degree (as it has in Texas and Northern Mexico), Blue Water II is used extensively for long drops. It has several major benefits for caving. First, it has no elongation or stretch at all. Any caver who has clipped into a line with his prusik slings or ascenders and found himself climbing 10 or 15 feet before he got off the ground will appreciate the non-stretch characteristics of Blue Water II.

Also Blue Water, like Edelrid and Mammut, is constructed with the core and sheath design so that it resists abrasion. This is very important when using mechanical ascenders which have teeth or grooves that bite into the rope to secure their grip. Bluewater does not twist when hanging free as do laid ropes. Again this is a great benefit to a caver, who otherwise may find himself spinning around at a relatively high velocity for several minutes at the bottom of a 500 foot pit. This may seem amusing to those watching on the top, but it is really debilitating and can be downright dangerous. The sheath construction of Blue Water was designed specifically to operate under the wet and muddy conditions frequently encountered in a cave and still allow the mechanical ascenders and prusik knots to grip the rope properly. Because of the fact that Blue Water doesn't stretch, the manufacturers specifically do not recommend it for dynamic belaying. For rappelling, the smooth surface of Blue Water (and the other core and sheath ropes) makes for a very uniform and somewhat faster rappel. For this reason, additional friction is needed for these ropes (see next chapter).

Blue Water II in its 7/16″ diameter has a breaking strength of about 7,000 lbs., considerably higher than any of the other climbing ropes available. Some cavers also consider Blue Water (and the other core and sheath ropes) to be safer, because the sheath protects the load bearing core of the rope and any damage that is done by ascenders or abrasion over unpadded rock is easily seen. Even if the sheath on Blue Water is cut completely through, 70% of its strength (4900 lbs.) still remains. On the contrary, laid ropes, because of their twisted construction, flex during use, and this can allow dirt to get between the strands, leading to possible wear.

Dacron and Other Synthetic Ropes

Dacron ropes, which are widely used in sailing in a laid construction, are not really suitable for caving because even though they have very little stretch, they are too soft for prusiking. Recently,

Polypropylene ropes have come into some use in caving and climbing for slings, either for direct use with prusik knots tied in them or as slings for mechanical ascenders. In general, these ropes can be quite suitable for this purpose in either the 5/16 or ⅜ inch diameter, in a laid or braided construction. Polypropylene ropes have a strength similar to nylon, so they test at approximately 2100 to 4500 lbs. in these smaller diameters. Tests have been run using Polyethylene for prusik slings, but Polypropylene seems to be more consistent in performance. The only word of caution in using any synthetic fiber on another synthetic fiber (such as a main climbing rope of either Goldline, Blue Water or Edelrid), is that if a very high speed were to be reached, the lines could theoretically fuse together. This is a very remote possibility, but worth mentioning if one is considering using several slings for hauling purposes where the speed might run fairly high.

CARE OF ROPES

Care in Use

Cavers and climbers have several strict do's and don'ts for ropes. The most common are never step or walk on a rope and never drag a rope across the ground or rock. Instead, pick it up in a coil or bunch and carry it. The reason for this is that small chips of rock and dirt can become lodged in the rope fibers and actually cut it. When not in actual use, a rope should be kept properly coiled or braided.

Immediately after purchasing a new rope, tape the ends carefully with two and a half to three inches of vinyl tape, stretching the tape carefully so that it conforms tightly to the shape of the rope. Then cut through both the tape and rope within about half an inch of the end to get a clean break. When cutting a longer rope into lengths, tape about a six-inch section and cut it in the middle so that both new ends are finished properly.

A better method of finishing the end of the rope is to similarly tape the rope tightly but leave about an inch of rope sticking out from the tape. Then fuse the end of the rope using a hot flame from either a carbide lamp or a stove so as to form the rope. It's best to use a spatula or similar tool when doing this and be very careful not to drip any hot nylon on the hands because it really burns; it has a tendency to stick to the skin and is very difficult to brush off. Also be careful that the fused end does not exceed the original diameter of the rope, otherwise there may be trouble in running it through a carabiner.

Besides protecting the end of a rope with vinyl tape some cavers mark the middle of the line or other check points as a convenience for judging how much rope has been paid out. Avoid, however, using any chemical or paint markers on rope which might cause deterioration. Some climbers have used fingernail polish successfully, but vinyl tape seems easier and safer.

Braiding a Rope

When carrying a rope in a cave, the standard coil of rope as used by climbers can sometimes cause a problem with the rope

Figure 7-2. Braiding a Rope. A braided rope is less likely to snag or tangle in caving usage.

snagging on projections. There are two good alternatives. The first is to carry the rope in a bag or gunny sack tied at the ends with parachute cord or similar line, perhaps with a carabiner inserted for hauling purposes.

The second method is called braiding, a technique which is used more commonly in Britain than in the United States, but which every beginning caver is urged to learn. Braiding is shown in Figure 7-2. Basically, what is involved is to double the rope several times, reducing its overall length to about 20-30 feet of doubled or quadrupled strands. Then after forming an overhand loop in the end, bring successive loops through each preceding loop as shown. By this technique it is possible to reduce a 120 foot length of rope to about eight or nine feet. This resulting braid is relatively flat and fits easily across the shoulders for carrying. Practice this technique at home along with knot tying and be prepared to volunteer for braiding or coiling duty, since these tasks usually devolve on a group's junior members.

Coiling A Rope

Another technique most new cavers can master at home is that of coiling. Several methods are in use today, but perhaps the simplest is the one where the rope is wound between the knee and foot as shown in Figure 7-3. The technique is first to run two or three loops around knee and heel and tie two overhand knots around them, leaving the short end of the rope about 12 to 15 inches long. Then the remaining rope is carefully wound between knee and foot, using the heel to prevent the rope from sliding off, until there are about ten feet left. At this point, thread the remaining length around and through the coil until the knot tied with the short end is reached. Here wrap several tight loops around the rope and tie off the end with a fisherman's knot to the remaining length of short end.

In general, although nylon is not affected by water, it's still not a good idea to coil a rope when it's wet. If this is necessary, it's best to re-coil it at the first opportunity, taking care to remove any kinks and leaving it spread out to dry before recoiling. After a particularly muddy caving trip, wash the entire rope in a washing machine using a mild detergent and either cold or lukewarm water. This can be done quite successfully when the rope is in either a coiled or braided condition, but it is best to stretch it out and air dry it afterwards. However, avoid lengthly exposure to the sunlight since nylon is somewhat subject to ultra-

(Photo by Charlie and Jo Larson.)

Figure 7-3. Coiling a rope.

violet deterioration. Also, avoid washing machines with glass doors on the front. At least one case has been reported of severe rope burning from friction with the glass during the spin cycle.

To protect any rope which must run over a sharp edge in a cave, some kind of padding must be used. This can be something as prosaic as a jacket or some 18 to 20-inch pieces of slit garden hose can be carried along to protect the rope in such contingencies.

Inspection for Damage

Inspect every inch of a caving rope before a trip or at least once a week during the heavy caving season. Figure on using one rope for only two or three caving seasons, then retire it from critical use. The British Mountaineering Council says retire it after 100 days of use.

With laid ropes, the amount of wear can be judged by checking the outside yarns of each strand. When these are worn through in an area more than one full turn of the rope (until only one third to one half of the area remains), the rope has lost about half its strength and should be discarded.

Ropes that have been subjected to severe elongation as in stopping a long fall during belaying, should also be retired from active service. Check for wear caused by severe shock by looking inside the strands. If a powdery fiber residue is visible the rope has probably been considerably stressed. Look also for any sections that look larger or smaller in their overall outer circumference than the ones nearby. Usually in a climbing rope, one complete spiral is about equal to the nominal circumference. If a turn is found that is considerably more than this in circumference or is as much as 25% less, it's probable that the rope has been badly strained. Chemical attack on a rope may show as a staining or softening of the fiber. Effects from heat are much more difficult to see. Excessive heat will cause fusing and glazing, but lesser heat such as from a carbide lamp can seriously weaken a rope without leaving any tell-tale signs.

With core and sheath ropes like Blue Water and Edelrid, damage is much easier to see, because it will be to the outer mantel which is specifically designed to take the wear, leaving the core undamaged. With Blue Water, even if the sheath is completely cut through, 70% of the original strength remains, although it should be discarded if this happens. Minor damage to the sheath can be repaired by wrapping vinyl tape around the rope to keep out dirt.

SLINGS AND SLING MATERIAL

Sling Materials

Today, slings are generally made up of synthetic fibers, although it wasn't too long ago that manila rope was widely used.

The beginning caver is well advised to standardize on one inch tubular nylon for all of his initial sling needs. Tubular nylon is very soft to the touch and tends to spread the load better than ropes when used in body and seat slings. For prusik knots, polypropylene is used. But, again, one inch nylon or Blue Water II are commonly specified for mechanical ascenders, and since ascenders seem to be replacing the prusik knot, it's better for the new caver to standardize on tubular nylon.

The Basic Sling

A universal sling that has a myriad of uses is easily made from an eight to ten foot length of white tubular nylon. One of the first uses this will be put to in any caver's life, is as a rappel sling. The rappel sling is about seven to seven and a half feet long and can be quickly made from the basic sling by simply tying a fisherman's knot at the proper length. The rappel sling is commonly called a diaper sling and is worn as shown in Figure 8-1 (next chapter). The second main use for the universal caver's sling is as a chest sling for a belay tie-in or prusiking. This is made up using the full length of the sling tied into a loop with a fisherman's knot, then drawing it over the shoulder and under the opposite arm as shown in Figure 5-5. The middle parts of the loop are brought together in front of the chest and tied in a sheet bend with the carabiner inserted in the resulting loop, as shown. A third use for this sling is to attach it to a handline or other fixed line with a prusik knot with the other end around the waist when working

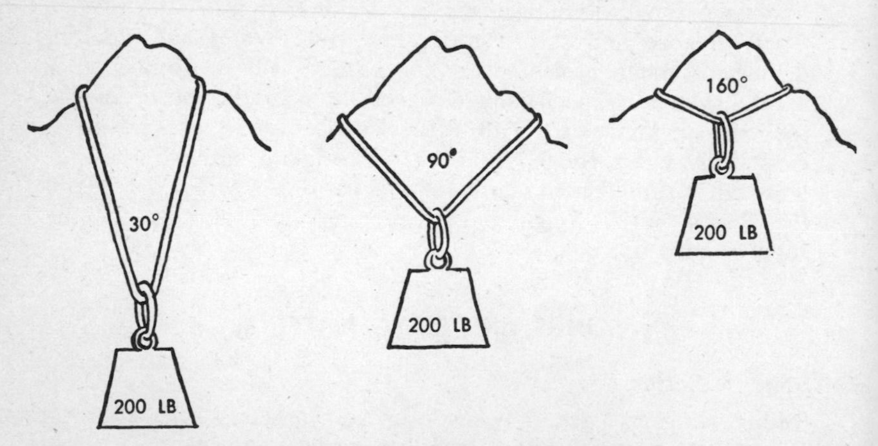

Figure 7-4. Loose Hanging Anchors. Slack anchors put much less stress on rope or slings. With a 30° angle, each side holds only 101 lbs. With a 90° angle, each side must hold 141 lbs, and with a tight 160° angle, the load increases to 575 lbs on each side.

in an area where there is some vertical exposure and when good safety practice calls for being tied into the rope.

Another use is as a hauling line to raise and lower equipment over short pitches. If each caver in the party has this universal sling, several lengths can be knotted together with fisherman's knots to provide a suitable length. Slings are also sometimes attached around rocks or massive formations to attach a ladder at the proper position in a pit or to thread a double rope through in order to pull the rope down after the last man is down. If used this way, the longer the sling, the more acute the angle obtainable between the two sides of the sling—highly desirable. For instance, with a 200 lb. load, a 30° angle on each side must hold only 101 lbs., but if a very short sling is used so that the angle is, say, 160°, each side must support 575 lbs.

HARDWARE

Carabiners

Carabiners are oval rings of special alloy aluminum or steel with a spring-loaded gate on one side. Present day carabiners test at about 3,000 to 4,000 lbs. and cost anywhere from $2 to $3. It is well to remember that a carabiner is less than half as strong when the gate is open as when it is closed. However, they are designed so that they can be opened under load and this is one of the characteristics important in rock climbing.

At a minimum, the caver just starting out should have three carabiners, two of the locking type which are inherently safer, and one of the standard oval type with a brake bar designed to fit that particular carabiner for rappelling. Carabiners have many uses in caving, including rappelling, snapping into a line when doing a horizontal traverse, attaching to a bowline knot tied in the end of a belay line, attaching gear to a figure eight loop tied in a lowering rope, and attaching equipment or clothing to the cave pack or line. Brand names available in the U.S. and Canada are SMC, REI, Bonaitei, Stubai, Cassiv, and Simond.

Bolts

Bolts and bolt hangers are used extensively by cavers to provide a fixed secure point at which to anchor ladders or climbing ropes. Bolts are preferred to the pitons used by climbers because it is

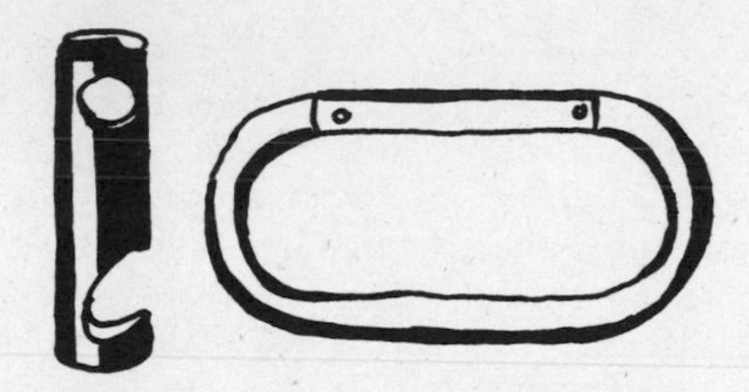

Figure 7-5. Carabiners. These are special high strength oval links with a gate in one side. A minimum of three are recommended: two with brake bars (center) for rappelling (attached together with a chain link), and one locking type (bottom). Several shapes are available. Breaking strength averages 3000 to 4000 lbs.

Figure 7-6. Rappelling Devices. Spools and rack at left, provide the varying amounts of friction needed for very long drops. Rack accepts up to seven brake bars. For average drops, double brake bar at right is the most common and is safer than a single bar. The two carabiners are held together with a chain link.

seldom that a suitable crack for piton driving can be found in the limestone of caves. Cave limestone is often crumbly and very breakable, or it may be covered with hard, brittle flowstone or calcide, which is similarly inhospitable for pitons.

The most common bolt sizes are ¼ inch but some cavers who like to pound use ⅜ inch. The ¼ inch tests at about 2000 lbs., the ⅜ inch at 3000 lbs. The strength requirements for bolts are about 2000 lbs. to support a belayed fall, 1000 lbs. for a rappel, and about 500 to rig a ladder or a hand line.

Bolts are installed using a hammer and an impact or star drill. In use, the star drill is struck repeatedly by the hammer and rotated slightly between each blow. It is also important that one blows the dust out of the hole as he goes along (but not into the eyes, which is easy to do) to make drilling easier. It usually takes 20 to 30 minutes to set a bolt, but it can take longer in especially hard limestone. Leave about ⅛ inch between the end of the thread and the hanger rather than tightening the hanger right up against the rock. This is so that the hanger can adjust itself to the movements of the load and not force the bolt loose when the load shifts. In setting any bolt, try to confine the damage to the cave wall to as small an area as possible.

The most common brand of bolt used in this country is the Rawl stud or contraction bolt. The shaft of the Rawl stud is split and offset slightly so that it jams into the hole.

Never use a bolt (or a piton if on a practice climb above ground) installed by someone else unless it was done by some member of the group or grotto and known to be regularly used and tested. An indication of the safeness of a bolt can be had by tapping it with a hammer to check the looseness. If there is any doubt, it is best to reset a new bolt. As with any type of anchor, artificial or natural, more than one is recommended when setting up a safety line or ladder, to act as a backup.

Climbing Wedges and Nuts

A very recent development in climbing hardware brought out by rock climbers is the jam nut wedge, which seems to now be replacing the piton. Cavers are just beginning to try them, so it's too early to tell if they will become as popular as bolts. Climbers for some time have been concerned about the litter that their fraternity leaves on well-traveled rock faces in popular mountaineering areas. It seems that nearly every possible crack has several pitons in it from someone's previous trip. Both the challenge of free climbing while placing one's own pitons as well as the ecological aspects have lead to the development of easily removable wedges and nuts. Figure 7-7 shows the typical configurations, which include various shapes including circular, hexentric, and wedge shaped.

Basically the function of the wedge is to act like a chock stone that has become forced into a crack or crevice. The advantage of wedges over either stones or pitons is that they are quite a bit safer. Usually they can be placed by just forcing them into position by hand, taking if necessary one or two taps to seat them firmly. They are also far easier to remove than pitons, which

Figure 7-7. Climbing Wedges and Nuts. These new climbing anchors test at 3000 to 6000 lbs. From left to right, they are Chouinard Hexentric, Clogwyn Chocks, Forrest Fox Head, SMC Wedge, and SMC Hex Nut.

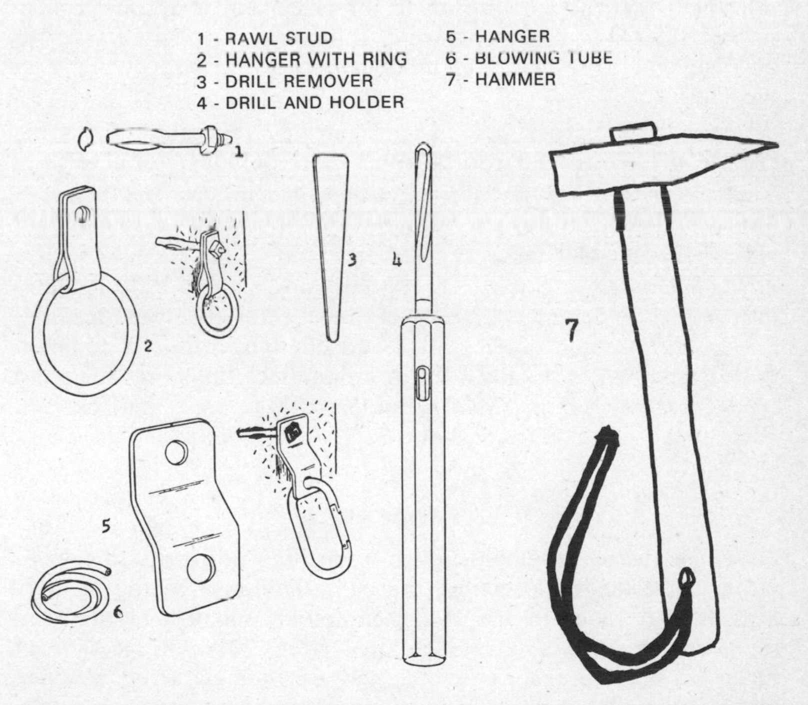

Figure 7-8. Bolts and Hangers. Bolts, like the nuts above, are used to provide anchors where no outcrop, boulder, tree, or stout formation is available. The quarter inch size tests about 2000 lbs. Tube is to blow dust from hole during drilling.

is the reason why so many pitons are left behind and are considered expendable by most climbers.

Basically a nut or wedge is made of metal or plastic resin in the form of a block in various shapes or sizes. It is inserted or jammed into the crack with some kind of sling attached. Several sizes are available and an assortment is needed just as an assortment of pitons is needed to fit the various cracks available.

Obviously great care must be taken in placing a nut or wedge to avoid its pulling loose or falling through if the crack is wider at the back or bells out unexpectedly just behind the points of contact. Generally, the nut should be placed as deeply as possible to reduce the load leverage on any thin or weak looking flake of rock. As with pitons or bolts, several nuts or wedges are preferred to just one whenever it's practical to do so. Two or more nuts of differing sizes can be placed in the same crack. However, in no case should one nut be jammed against another nut. Also remember that the nuts are made for placement in vertical cracks so that the load pulls them firmly into a tighter position. Avoid any use where the pull will tend to unseat the wedge or nut in the crack.

Most nuts have a hole drilled through the center the same size as the common types of ropes and webbing, running anywhere from ¾" to one inch. The one inch tubular webbing recommended for the novice caver will fit into most of the wedges and nuts available.

Test strength of these new devices ranges from 3000 to 6500 lbs., which is actually more than most carabiners and matches closely the strength of the slings and climbing ropes in common use. Brand names usually seen in mountaineering stores for these types of devices are SMC, Chouinard, Clogwyn, and Forrest. They range in price from about $1.25 to $2.00.

LADDERS

Although recent developments in rappelling and prusiking techniques have tended to eclipse the cable ladder in caving, ladders are still very much in use. Most commonly today, they are used for climbing out on pitches ranging from 20 to 40 feet or at most, 60 feet. In many cases the way down is via a rappel, since this is quite a bit faster and in many respects is safer than climbing down a ladder, which is somewhat more difficult (although easily mastered) than climbing up a ladder.

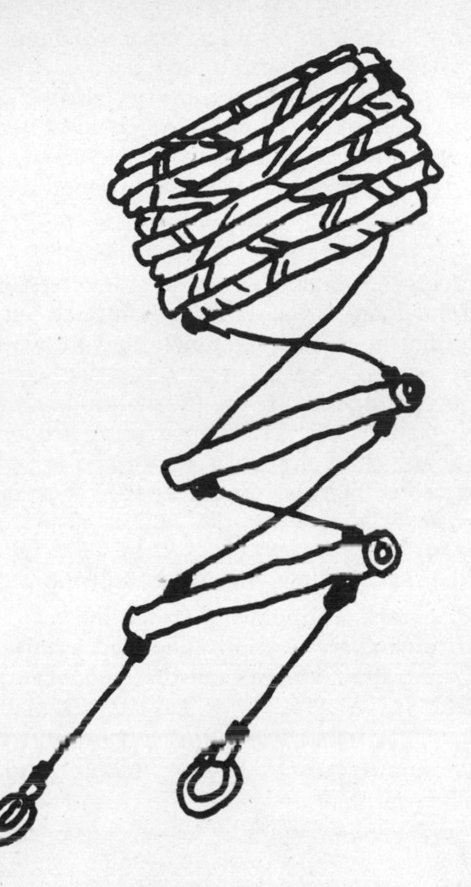

Figure 7-9. Steel Cable Ladder. For storage, ladder is rolled as shown. Ladders test at about 650 lbs, far less than other climbing gear, so a dynamic belay is always used.

A common method among American and Canadian cavers today, is to arrive at a pit, rig a ladder and a rappel line at the same time, then rappel down. On the way out, the rappel line is used by the first man with some type of prusik self-belay as he climbs the ladder, then he belays the rest of the party as they climb the ladder.

Types of Ladders

Modern caving has now completely standardized on the cable ladder and it is seldom that the rope ladder with wooden, metal, or rubber rungs is seen at all. The principal problem with rope ladders is that they were made of nylon rope for safety. (This was before polypropylene and dacron ropes arrived on the scene.) Since nylon stretches so much, however, a caver had to climb the first eight or ten feet twice before he got started off the floor just to take up the slack in the nylon rope. For this reason cable ladders, although considerably more expensive to buy or construct, have replaced rope ladders completely.

Cable ladders are made of high grade aircraft cable with rungs of duraluminum. The ends are finished off in Brummel hooks for attachment to two carabiners and thence to an anchor. Cable ladders normally come in a ten meter length (33 feet), with a spacing of about 12 to 15 inches between rungs. In Britain, where cable ladders are used more extensively than in this country, many clubs make their own and are careful to keep the rungs no more than 10 inches apart. It is far more comfortable to climb a ladder with the smaller distance between the rungs. The rungs themselves are 5¼ inches long which makes them more than adequate for one boot, but too tight for two.

Cable ladders test at about 1000 lbs. for each of the cables, and 650 lbs. minimum for each rung and clamp that attaches the rungs to the cable. This means that the effective strength of the cable ladder is 650 lbs. Some ladders are made of stainless aircraft cable which makes them a little more durable in use.

The ladders most commonly sold in this country today are made by a French manufacturer, Pierre Allain, and they are available through caving suppliers shown in the appendix. They run about $1 a foot or $33 for a ten meter length. Many clubs have two ladders, but seldom more, because sixty feet is about the maximum ladder that anyone climbs these days. Pitches longer than that are usually easier to surmount using ascending devices and a fixed rope.

All cavers, including new cavers, should be especially wary of older cable ladders, particularly those made in the 1950's and early 1960's. Some of these ladders had a problem with electrolysis taking place between the different types of metal used in the rungs, the clips, and the cable, which caused the cables to break at those points. But even a brand new ladder in perfect condition still presents the possibility of danger so every club in the United States, Canada, and Britain now has a standing rule that ladder climbs are belayed with no exceptions. Even a short 10 to 15 foot climb is dangerous without a belay.

Care of Ladders

In the cave environment, protect a ladder from damage and abrasion by carrying it in a cloth sack, cave pack, or rubber innertube. When installing the ladder down a pitch, don't throw it over the end but lower it gently. Also don't stand on a ladder stretched out on the ground—it's the same principle as never standing on a climbing rope.

After use, wash all the mud off the ladder and inspect it when it's dry. Before using the ladder, it should be tested by having three men pull on it strenuously with the ladder attached to a tree or the bumper of a car. When storing the ladder between uses, coil it up relatively tightly, twisting each of the rungs in the opposite direction to form a crisscross pattern in the cable. However, when storing a ladder during the off season it is better to uncoil it and keep it off the floor in a dry, airy place; do not store it tightly coiled.

SCALING POLES

Poles are not much used in American caving, but they have been used extensively in European caves. Basically a scaling pole is a device with which to raise a ladder or in some cases a line that can be climbed to reach a passage high up on the side of a wall. They are usually made of sections of aluminum tubing attached together with some kind of bolt or screw system. Guy wires are attached to both sides for stability. Generally the pole is carried into the cave unassembled, then put together in position and raised to the lip or ledge with two or three feet extending over the top so the climber can easily climb off.

WET SUITS, DRY SUITS AND BOATS

To push wet passages in caves with streams or lakes, the wet suit has come into common use both in this country and in Europe. A wet suit is made of neoprene foam rubber and is worn directly next to the skin. When it is dry, its thousands of tiny air pockets trap air which is warmed by body heat and insulates the caver better than clothes can do. Then when the rubber gets wet, water flows into the tiny pockets in the foam, and is quickly raised to body temperature, providing an equally good insulating layer that keeps the caver warm. This kind of suit is called a wet suit because it doesn't really keep the wearer dry; it keeps him wet, but warm. Usually these are made in two-piece construction. For caving use they should be purchased a little looser than for scuba diving.

Dry suits are of thin rubber construction and designed to fit closely to the body so as not to allow any water to get beneath the suit. However, they tend to tear quite easily and so are not as suitable for caving purposes as the wet suit, even though they may be available in some areas at a bargain price in surplus stores.

Wet suits are always worn with normal caving clothing or at least coveralls over them to prevent their snagging and tearing. The ideal fit for a wet suit is such that it should just touch the body at all points when the wearer lifts his arms over the head. The best size for wet suit material is 3/16 of an inch (6 mm). For caving uses, since a wet suit tends to be worn for a much longer period than when diving, the two-piece version with a nylon lining is more comfortable as well as far easier to put on.

The cost of a wet suit today runs about $25 on up.

MECHANICAL ASCENDERS

For prusiking purposes, two types of mechanical ascenders have nearly replaced the prusik knot. These are the Jumar ascender and the Gibbs ascender. Basically, they are devices that can be easily moved up the rope, but when the upward motion is stopped,

Figure 7-10. Mechanical Ascenders. Gibbs on left, Jumar on right, grip the rope firmly with weight applied, but release easily when moved upward.

(Photo by Charlie and Jo Larson.)

Coming out of the 60 foot entrance pit in Big Horn Caverns (Montana-Wyoming border). One caver is climbing the rope using Jumar ascenders. Another is climbing a cable ladder, with a belay rope attached to her body for safety.

they grip the rope securely and will not slip. This is the same principle as a prusik knot. The advantages and disadvantages of the two types will be discussed in detail in the next chapter.

RAPPEL RACKS

In addition to the brake bar inserted across the carabiner, caving has developed some specialized rapelling devices for use in very deep pits. These are called rappel racks or rappel spools. (See Figure 7-6.) Basically, the problem is that when making a very long drop the length of rope hanging below the caver increases the friction between the rope and the rapelling device to the point where the caver may not be able to move at all. In some cases he may even have to almost force the rope through the rappel device. Then, when near the bottom, friction is less. Clearly a variable friction device is needed, hence the development of rappel racks and spools.

Rappel racks are essentially a long carabiner-like device having multiple brake bars. When the caver starts a long descent he may use one or two brake bars. Then at several points as he descends, he can add another brake bar completely safely, by pulling up the rope from below and snapping a new bar into place. This reduces his speed as he nears the bottom so that the entire descent is made very smoothly.

The principle of the rappel spool is similar. It depends on taking only one or two turns around it when starting out and adding several more turns as the caver descends. The principle again is to have very little friction at the start of the drop and more near the bottom.

These devices find their use in pits of 100 feet or more and work equally well in pits up to 1000 or more feet.

8

Vertical and Other Advanced Techniques

On this side of the Atlantic, the decade of the 60's saw a tremendous surge of interest in vertical caving coupled with the development of several new techniques and pieces of specialized vertical equipment. Without doubt, these developments have opened up many new areas of existing caves and led to the discovery of several new caves that could not have been explored with earlier available techniques.

Today, vertical caving techniques are used by nearly all cavers to some extent, but there still remains a particular breed of vertical cavers who delight in pits ranging in depth from at least 150 feet and going down to 1000 feet or more. Sometimes they are referred to affectionately by other cavers as vertical freaks, but there is no doubt that the techniques and equipment they have developed have benefited the entire caving fraternity. In particular, the development of new rappelling and prusiking techniques by Alabama and Texas cavers, primarily, have changed the character of caving in the U.S., Canada, and Mexico.

RAPPELLING*

Rappelling is a mountaineering technique that has been used for many years to descend a pitch rather than climbing back down. The difference between the goals and needs of mountain climbing and caving should probably be mentioned here again. Basically, a climber is interested in climbing a rock face or mountain. Many times he will not descend by the same route at all. In fact, he may walk down a very easy path on the other side or if this is not possible, will likely rappel down, since climbing down is much harder than climbing up. A caver, on the other hand, usually starts his climbing by going down, since in general most caves lead downward rather than upward. For that reason rapelling is one of the very first things that a caver does in most any cave with substantial vertical extent. Rappelling is thus a primary technique with cavers rather than a somewhat secondary technique as in rock climbing.

Fundamentally, rappelling is a method of sliding down a rope safely. Obviously this cannot be done just with the hands for more than a few feet, so climbers first developed a technique which cavers irreverently refer to as the "hot seat rappel," otherwise known as the body rappel. This method won't even be described here since the author has never seen a caver use it.

Instead of the body rappel, cavers rappel with a carabiner, brake bar, and a diaper sling. The sling is made from a piece of one inch tubular or flat nylon, or from standard climbing rope if the more comfortable sling material is not available. Total length is about 7½ to 8 feet. It is attached to the body like a diaper as shown in Figure 8-1, and the three resulting loops from the diaper configuration are attached together with a carabiner. This should be a *locking* carabiner since its superior margin of safety is well worth it. The carabiner is first attached to the sling with the gate down and facing the body since it's easier to put on this way. Then after all three loops are inside the carabiner, it is rotated 90° or one-half turn so that the gate is up and away from the body. In this position it won't push into the body or accidentally open. Even though a locking carabiner is used, this is still more comfortable since it prevents the threads of the locking device from pushing into the body. The key words are "up and away" in remembering the final position of the gate on the carabiner.

*Rappelling is called *abseiling* in Britain. Rappelling is the French word; abseiling is the German word.

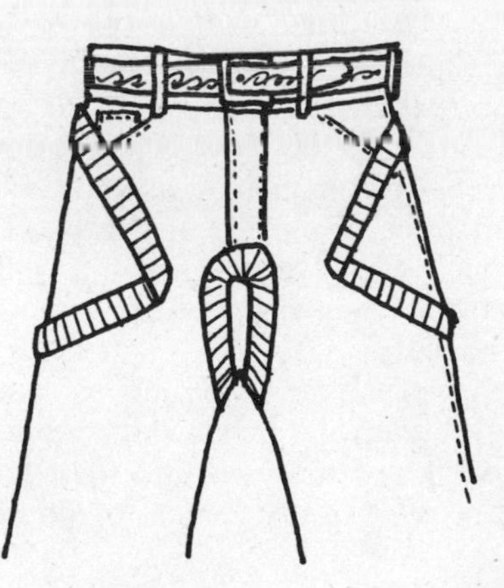

Figure 8-1. Diaper Sling for Rappelling. Made from Universal Caver's Sling. (See Figure 5-5). It is attached in the front with a locking carabiner.

The rappelling carabiners and brake bars are then hooked into this locking carabiner, with the brake bars in the open position. The caver first stands with the rappel rope between his legs, then reaches down, picks it up, and places a loop of rope up through his carabiner. The brake bar is then snapped into position and the slack taken up between the anchor point and the caver's position. Then the second carabiner/brake bar secured to the first with a chain link, is attached to the rope in the same way. Remember, all carabiners must have the gates up and away from the body. For very long rappels, the rappel rack described in the previous chapter is commonly used in order to provide a variable amount of tension via its multiple brake bar system to counteract the weight of the rope trailing below the rappeller.

In all cases, getting into the rig should be done well back from the edge of the drop or pit for safety's sake. Be sure to recheck the rig at least twice before going down: the first time immediately after the rappel line is inserted into the carabiners, and the second time just before stepping out over the edge. It's quite easy for the rappel line to become loose or to spring out of the bar when one is moving around at the lip of a drop. Pressure on the rope during the rappel itself keeps everything in place, but it's also smart to check the rig several times later in the rappel, particularly if stops are made on any ledges.

From the brake bar, the caver wraps the rappel rope around his right hip, grasping the rope tightly in the right hand. This right hand is the braking hand and must never let go of the rappel rope. To assure that the rope is not dropped, the use of the rappeler's bracelet is recommended as shown in Figure 8-2. This eight to ten inch piece of sling is tied around the wrist, and has a carabiner through which the rappel line is threaded. This prevents the rope from being dropped in case of an accident or a sudden jar. The left hand rides in front of the rappel rope, slightly above—at chest or head height—and is used for balance, but not for control. Downward motion can be stopped instantly by bringing pressure with the right hand against the hip. Gloves must be worn on both hands, of course, to prevent rope burn.

From the novice's standpoint, the only really tricky part of rappelling is that first step over the edge of the lip. For this reason, all beginners should be belayed on their first few rappels even though most cavers agree today that a belay is more trouble

Figure 8-2. Basic Rappelling Position. With feet spread apart and weight borne by the rope, you literally walk backwards down the face. A quick stop can be made by simply bringing the right hand tightly against the hip. When rappelling free (away from a wall), shift position slightly so you are sitting upright with the legs extended almost straight out. Note bracelet of sling and carabiner to prevent dropping rappel line.

than it is worth when rappel technique has been learned properly. But for the first rappels of a beginner, a belay is necessary.

Stepping out over the edge is done as follows: With the rope taut between anchor point and rappel sling, the caver steps backwards with his legs spread as far apart as possible for comfort and balance. He then literally walks down backwards with his feet firmly planted on the rock face. In caving, the rock face frequently drops back sharply or bells out as in a pit so that the caver's feet can suddenly lose contact with the rock face. Aside from the initial shock, this really creates no great problem, since all he has to do is lower his feet slightly and sit more upright, rather than in the crouched position used when the feet are touching the wall.

The novice will find that he will move so slowly at the beginning of a rappel that his problem will often be to push the rope through the carabiner with the right hand. This tendency should be avoided because it removes the right hand from its normal braking position behind the right hip. However, if pushing the rope really is necessary, be sure not to drop the line. Return the hand to its braking position as soon as possible.

One thing the beginner should keep in mind is that an even, steady descent is essential so as not to put any undue tension on the rope such as caused by a sudden braking after a long free slide down the rope. The proper technique is to push off slowly and gently, keeping the legs apart for good balance, but not trying to win any speed contests on the way down.

Rather than a separate belay, once the beginner has learned the technique of rappelling, a very effective belay can be run from below by having another caver hold the bottom of the rope. In case of an accident, the caver below can stop the descent by simply pulling the rope taut. The natural straightening of the rope increases the tension over the brake bar and will bring a falling rappeller to a safe stop quite easily. However, the caver below should keep in mind that he must not stand directly below the rappeller because of the danger of falling rocks dislodged during the descent.

In mountain climbing, it's very common to double the rappel rope, pulling it down after the descent has been made. However, in caving the first direction is usually down rather than up, so rappel lines tend to be left in place for use on the way out for either prusiking or as a belay line for a ladder climb.

Hasty Rappel

For very short pitches of 10 to 20 feet the technique of the hasty rappel is very handy in a cave. Basically the caver runs the rope across his shoulders and extends his arms as far as possible. Then he grasps the rope in each hand after taking a turn or two around each wrist. This can be done facing either to the right or the left depending on which way the slope runs and which way visibility is best. The principle is very simple. The caver just walks down sideways keeping his head away from the body in order to see where he's going. To halt the descent, the lower hand is simply brought up to the chest and all motion stops immediately.

Practice Climbs

As is the case with all climbing techniques, one usually learns the basics at a practice climb above ground with a group of experienced cavers who can provide the fine points of technique as well as the advice and encouragement needed. Before attempting any 300 foot pits, a caver should have rapelled extensively in dozens of drops so that the technique is completely second nature.

LADDERS

Climbing

As mentioned before, most descents in caves are made by rappelling. This is true even though the way out may be by a previously rigged ladder, since ladders are somewhat more difficult to climb down than up and rappelling is faster and easier. But the cable ladder is still very much in use for coming out of a cave, because in general it's faster (and cheaper) to learn ladder climbing than prusiking, it doesn't damage ropes the way mechanical ascenders do, and it doesn't require each caver to have his own custom-fitted prusik slings and harnesses. Despite the fact that a caver, once rigged and on the rope, can move faster and with less effort by prusiking, if the entire group isn't experienced and individually equipped with prusik gear, it is usually faster to exit via ladder.

Any caver can master ladder climbing on his second or third try. The only trick is to place one foot behind the ladder on every other rung to keep the feet from pushing out from underneath. To get started, most cavers face the ladder and place the first foot

Figure 8-3. Hasty or Arm Rappel. Very useful for short 10 to 20 foot drops. Rappel line can be made from two or more Universal Caver's Slings.

Figure 8-4. Climbing a Free Hanging Ladder. One foot goes on front, the other on rear. Hands can also alternate, or both go on back. Always belay every ladder climb, no matter how short. Ladders test at only 650 lbs.

(usually the left) behind the ladder on the bottom rung, reaching back through with the boot to hook the heel on the rung. Then they bring the left hand around behind the ladder at a comfortable height on a rung above the head, and shift the weight up onto the left foot. Next the right foot is placed on the second rung, this time on the front of the ladder, the right hand going above the left hand, again on the front side, and so on to the top. Remember, climbing a ladder is strenuous (as any house painter or chimney sweep will confirm), so climb with the legs, not the arms. Walking up stairs is a lot easier than doing pullups. The human body is made with legs far stronger than arms so let the legs do the work. Use the arms for balance.

This technique of alternating the feet and arms on the front and back of the ladder works very well when it's hanging free, but isn't usually desirable (although it can be done) when the ladder is pressed tightly against an overhang or wall. In this case, the problem is to force the ladder away from the wall, so the hands and feet can get a good grip on it. The easiest way to do this is to turn the ladder slightly on its side, which takes a bit of strength, but is easy to maintain once done. Then place both feet on the front side, rather than alternating front and back, and push the ladder away from the wall with the tocs of the boots. It also helps to bend the knees and waist more than usual when negotiating an overhang or wall.

Although these techniques are easily learned, climbing a ladder is still inherently dangerous, particularly with wet muddy hands and when the fatigue factor comes into the picture. For these reasons, plus the fact that ladders do break, never climb a ladder without a belay no matter how short the pitch is. When all a caver's strength is concentrated on his feet and hands climbing a muddy ladder, he needs the benefit of a separate safety line around his chest with a fellow caver above belaying him.

The first one up a ladder will often climb using a prusik self belay from a waist or chest sling, tying into the rappel line with either a prusik knot, a Jumar, or a Gibbs ascender. In this usage, a Gibbs ascender is preferred since it will follow up behind with no effort required to move it along. On the contrary, both the prusik knot and the Jumar ascender must be loosened and moved up and this requires hanging on to the ladder with only one hand, a tenuous proposition, in any case.

When the last man comes up a ladder, it is sometimes a good idea to attach the bottom of the ladder to his chest sling with a short length of rope. This way the ladder will follow him up as

Figure 8-5. Climbing Close to a Wall or at an Overhang. Push ladder away from wall with toe of boots, or rotate ladder 90°.

he climbs and he can unsnag it from any overhangs or rocks, making its removal far simpler.

Rigging

When rigging a ladder be sure to select an anchor point that allows two or three rungs of the ladder to remain above the lip. This makes it far simpler to get off the ladder when the caver reaches the top. Surprisingly enough, if the ladder is not rigged in this way, it takes a great deal of strength and an inherently unsafe lunge to make that last three or four feet up over the ledge. The same is true for rigging a belay or prusik line. Rig it well back from the lip of the drop.

When lowering a ladder, don't throw it down, but let it pay out slowly so it doesn't become snarled on any overhangs. It's also wise, if possible to do safely, to watch it as it goes down so it doesn't route itself between any overhangs or through any holes that a cavers body couldn't fit through. It's the responsibility of the first man down to check the routing of the ladder. If necessary to reroute it, he should be given a very tight belay so he can rest and move the ladder safely.

ROPE CLIMBING AIDS

Prusik Knots

The original technique of prusiking was first developed in Austria by Karl Prusik and written up in the Austrian Alpine Journal in 1931. It was primarily seen at that time as a means of self-rescue for a climber who had fallen into a crevice.

Today, the prusik knot is still the one most commonly used (by those who use knots instead of mechanical ascenders), even though several other knots have come and gone. In fact, hardly a caving season goes by without someone developing a new knot which purports to improve the basic prusik knot. These won't be described here because the prusik knot, itself, is still quite adequate.

Regardless of the way that a caver attaches himself to the prusik rope, whether by prusik knots or mechanical ascenders, the technique is basically the same. Either two or three slings are used. The three sling method is the oldest and will be de-

Figure 8-6. Rigging Ladder. Leave two or three rungs above edge of drop, making it easier to climb off. The same is true of belay and prusik lines. Tie them off back from the lip so cavers can derig safely.

scribed to explain the basic movements. Separate slings are attached to each foot and to the chest, and these are attached to the rope with a knot or ascending device. The weight is supported by the foot slings. The chest sling, and the arms are for balance. To secure the slings to the feet, a snug loop is used in the bottom of the sling, and often some auxiliary cord or large rubber band (cut from a car innertube) is run around the ankle to prevent them from slipping off.

The technique is to stand astride the rope, raise the chest knot or ascender as high as possible, followed by one of the foot slings, usually the left, but always the one attached to the middle knot, in any event. Then the weight is shifted upward, so that the left heel is positioned under the body. Next, the right sling is raised to put the weight on both heels, and finally the caver stands up on both feet, sliding the chest sling as high as possible at the same time to start a new cycle. Since he is tied into the rope at three points, a separate belay line is not necessary.

From the beginning, it was obvious that for any but the shortest pitches, some way would have to be devised to get rid of the third knot. Several techniques in use today do this. One of these simply runs the two foot loops through a carabiner attached to a chest sling and thence to the rope itself. This means that the third knot is eliminated but there is still a good balance point because the slings are physically close to the chest. Climbing with either the two or three sling method can be tiring but a steady rhythm with frequent short rests makes it possible to climb thirty or forty feet a minute with minimum training and practice.

One critical thing with any kind of prusik arrangement is the length of the slings. Depending on the method chosen, the length of the slings differs somewhat as it does for each caver. It is very seldom that one caver can successfully use another caver's slings perfectly without some discomfort.

Several methods have been devised to take some of the pressure off of the chest and to provide a rest position for very long rappels. Chest slings (Figure 8-10) which circle the chest and go up either over one or two shoulders have been devised by enterprising cavers. Usually these are attached to the front by a carabiner through which the slings are run or attached. Similarly, elaborate seat slings have been devised, usually still used with a chest sling because although the seat sling is excellent for resting, it places the balance a little low and a routing of the slings

Figure 8-7. Basic Prusik Technique with knots.

through the chest is still desirable. With either a chest sling or a combined seat and chest sling, most of the weight of the body is supported by these slings rather than by the feet when resting, although the legs still do the climbing. One problem with a seat sling is that sitting down lowers the climber several inches, which must be made up when he stands up again.

Jumar Ascenders

Jumar ascenders are sold in pairs at a price of about $34. Even with this price, they have become extremely popular in caving and in some climbing groups, too. Jumars are used with a chest sling and carabiner for balance, through which the sling ropes are routed from the ascenders to the feet. Immediately after their purchase, most cavers replace the cotton sling rope supplied with the ascenders with nylon, usually, one inch tubular sling material polypropylene, or 5/16 Blue Water II.

Proper technique is as follows. To climb, a caver raises the lead foot, which is the foot attached to the highest Jumar. Normally this should be the right foot, since it is going to end up doing most of the work. After raising the right foot and its corresponding Jumar, the left or trailing foot and its Jumar are raised. With this technique, the legs support the entire body and there is little or no weight on the chest or arms. However, the arms do get tired, since they are raised above the head—an inherently tiring position. In general, small steps of 15 to 30 inches are most common and easier.

One problem with the Jumar system is that there is no real rest position. Several alternate methods have been devised using seat slings and a third Jumar to provide a rest position. However, these are best learned in practice with experienced cavers and since they differ quite a bit in detail, they won't be described here.

Negotiating Overhangs

Getting up over a lip or overhang can be a problem with any prusik method. Most cavers seem to prefer pretty much the brute force method in which they simply raise the prusik knots or ascenders as high as possible and try to force the rope away from the rock by pressing out with the feet. Then they slide the hand under the rope to move the knot or ascender past the tight

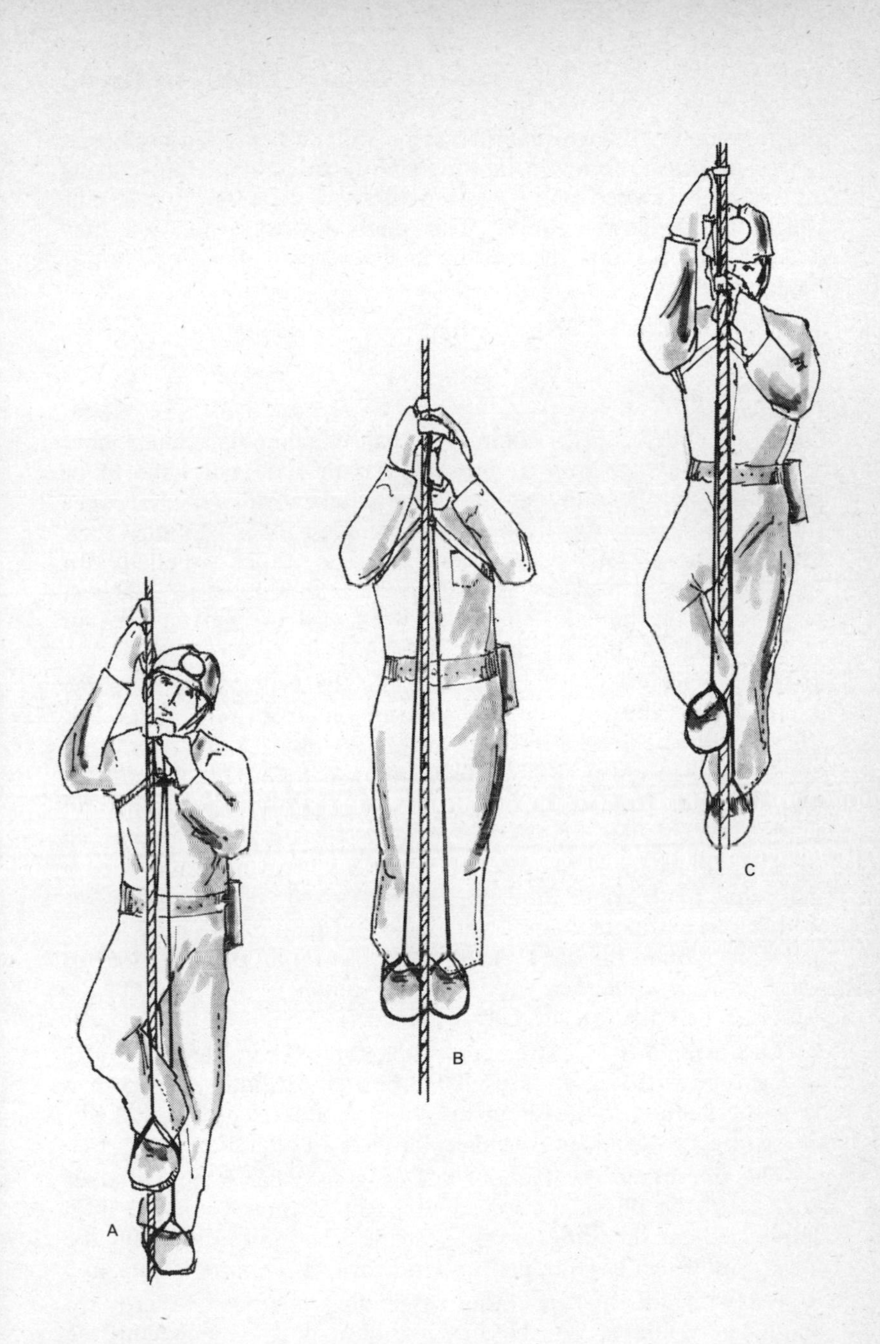

Figure 8-8. Jumar Technique. Slings attached to feet run through chest sling carabiner on way to ascender. Each foot and Jumar ascender is raised alternately. Chest sling provides balance.

spot. In really tough situations, some will carry a third ascending device or prusik sling which they simply tie above the overhang and carefully switch one foot into the new ascender to take the weight off the lower slings. This method works well, but may be more trouble than its worth, unless several overhangs are to be encountered.

Gibbs Ascenders

Gibbs ascenders, also commonly called climbing cams, have a major advantage over Jumars in that they do not have to be raised but simply follow along as the climber walks up the rope. In fact, the technique itself is described as rope walking rather than climbing. Gibbs ascenders are less expensive than the Jumars, being priced at $14.50 to $17.00 a pair (1972) depending on the model. They also work well on wet, muddy, or icy rope, which Jumars don't.

Gibbs ascenders are unquestionably the fastest way to climb a rope. In 1969, at the NSS convention in Lovell, Wyoming, Charles Gibbs climbed 100 feet in 43 seconds. This was bettered by a Canadian caver, Keith MacGregor, at the 1971 convention, who did 100 feet in 35.5 seconds using a somewhat different sling arrangement.

The Gibbs method uses two ascenders, one attached with a short sling to the right foot, the other attached with a longer sling so that it is positioned at the left knee and held in place by a second strap around the knee. A third strap is run around the chest or waist for balance either with the rope simply running through a carabiner or with a third Gibbs ascender.

The technique is extremely simple, the caver literally walks up the rope with the arms used for balance. Almost anyone can do it successfully if the slings are of the proper length and tightness and good balance is attained through a chest sling.

The important point about a Gibbs ascender is that it does not have to be physically raised. Instead, it moves upward by a simple pull on the sling or rope, yet it grabs instantly when the upward movement is stopped or a downward pressure is put on.

As with all of the other ascending methods, there are countless variations on the basic Gibbs sling arrangement too numerous to detail here and they must be tested by each individual, in any event.

Figure 8-9. Gibbs Technique. Three-Gibbs rigging with seat sling and shoulder-mounted third ascender. Caver simply walks up the rope with Gibbs method.

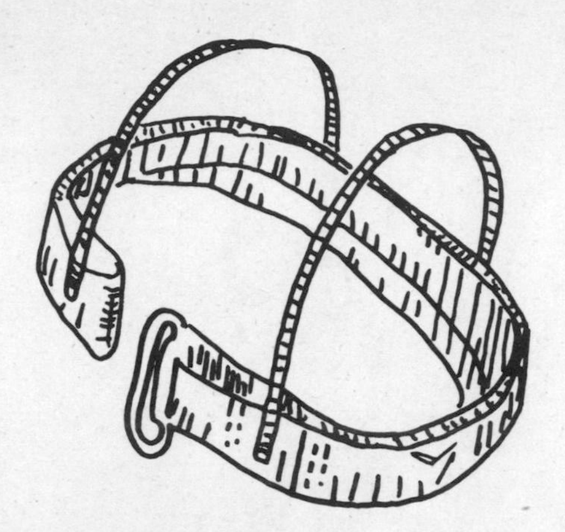

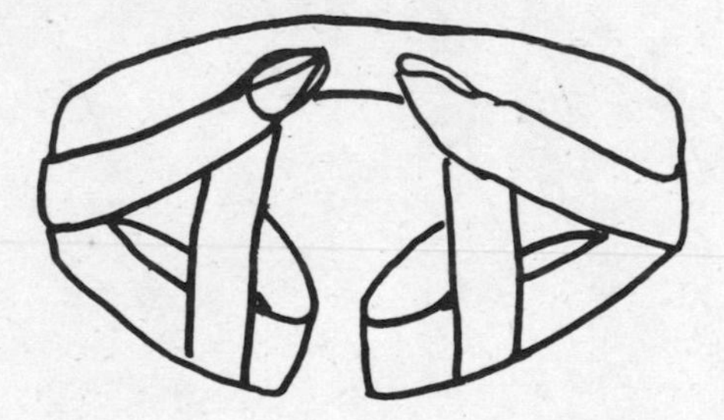

Figure 8-10. Chest and Seat Slings. Provide balance and resting position when prusiking. Top sling is made from two or two-and--one-half flat nylon (seat belt material or similar). The bottom one is made from one inch tubular nylon.

The Jumar/Gibbs Debate

The debate continues between proponents of the two prusiking methods, Jumar and Gibbs. The main advantages of the Jumar system is that it is more versatile since it can be used as easily to descend as to ascend. That is, by removing the downward pressure from the sling, the Jumar catch can be released and as easily moved downward as upward. Now, this doesn't mean that cavers have taken to rappelling with Jumar ascenders, however, there are situations at overhangs and outcroppings where it is sometimes better or more convenient to go back down a few feet and either readjust the rigging or choose another route.

Gibbs ascenders, on the other hand, are much faster and work on muddy or icy rope, as well as being cheaper and stronger than the Jumars. Gibbs test up to 2200 lbs., but they are advertised as being tested to 1000 lbs. Jumars on the other hand have been tested to about 700 lbs. The major disadvantage of Gibbs ascenders is that they are harder to remove from the rope and one cannot descend with them. Although some cavers seem to have devised methods for descending with the Gibbs, they take a great deal of practice and are not considered very practical methods.

Gibbs ascenders can be rigged on the shoulder. So rigged, they tend to keep the caver more vertical and this reduces the effort involved. In contrast, Jumars are used with chest slings or chest boxes of varying designs which keep the rope away from the body somewhat. The consensus seems to be that the Gibbs are superior if all drops to be climbed are long and free, but that the Jumars are probably more versatile because they can be used with several different types of systems and are easily clipped on and off the line as needed.

Texas cavers, who with the Alabama and other Southeastern cavers have pioneered many of the vertical methods being used today, report that a combined system is now becoming popular. This is particularly used in the deep pits in northern Mexico where the Texas cavers do a good bit of their vertical work. With this system a Gibbs is used on one foot, a Jumar on the opposite knee and a Jumar on a chest harness/seat sling. This seems to give the advantages of all systems.

As for the old fashioned prusik knots, they are still very much in use today, particularly for shorter drops where the added complexity of rigging into the line with ascenders is too much trouble. Also when equipment weight and bulk looms large or where long backpacking trips call for minimum equipment, prusiks may be taken along in preference to the heavier Jumars or Gibbs.

(The references in Chapter 11 cover several detailed expositions of prusiking techniques in common use in the National Speleological Society today, for those interested in pursuing the matter further.)

BELAYS

Dynamic Belaying

Dynamic belaying* was developed by climbers in the late 1940's to protect the lead climber who normally is climbing above the others. This is called an upper belay, because the climber is above the belayer. The practice has been adopted almost without change for caving with the exception that in caving the most common belay is a lower belay, it being more common for the climber to be below the belayer, although both upper and lower belays are found in both.

The principle of the dynamic belay is that if a climber falls, he is gently arrested by the belay rope and belayer rather than coming to a sudden stop, as would happen if he was tied into only a fixed rope. Factors that make a slower stop possible include the inherent tension around the belayer's waist coupled with the natural reaction time he needs to bring the rope to a stop. Thus, the belayer's body acts as a combination shock absorber and friction brake.

The positions for upper and lower belays are shown in Figure 8-11 and 8-12. With the lower belay, the sitting position is safest and best. The belayer sits with his legs spread apart somewhat, his knees flexed but not locked, and his feet against a tree (if at the entrance to a cave) or against some other solid surface like a cave wall. If there is any doubt about the security of the belayer or any possibility that he might be pulled from his position in case of a fall by the climber, he should be tied in to a firm anchor. When tying a belayer into an anchor point, there should be no slack in the tie-in rope.

The belayer runs the rope around his stomach or chest, the end nearest the climber held normally by the left hand which is called the feeling hand. The rope then passes around his back into the right hand which is called the braking hand. A belayer *never* lets go of the rope with his braking hand under any circumstances, but instead lets it glide slowly through his clenched hands. When gradually paying out line as the climber descends, the belayer tries to keep the rope as snug as possible, but also must develop a feel for the rate of climb so that the climber isn't held back or slowed down in any way.

---

* In British usage, the term belay means a fixed anchor point to which a rope or ladder is attached. Thus the British can have a belay point or anchor, just as is done in the U.S. and Canada, except when the North American caver says, "belay anchor," he only means an anchor for a belay, not one for anything else. The British call dynamic belaying "life lining" and the belay line a "life line."

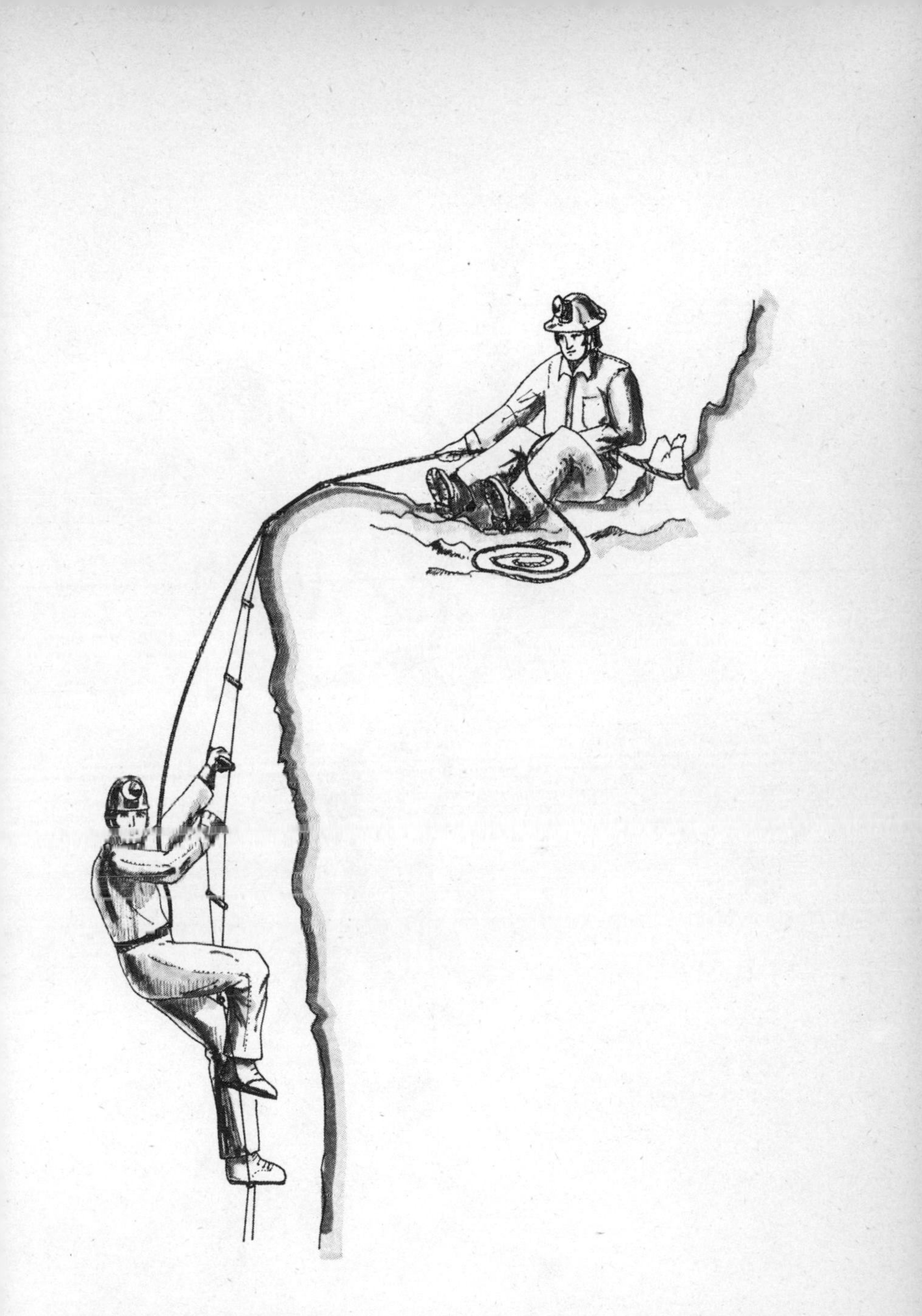

Figure 8-11. Sitting Belay Position. Used for a lower belay (climber below). Note position of coiled rope at side of belayer, and tie in rope behind belayer.

Figure 8-12. Upper Belay Position. Rope goes under seat of standing belayer.

To stop a fall, the braking hand seizes the rope and brings it quickly across the chest. The beginner must remember to keep his arms fully extended as he pays out the line so as to get maximum braking power when he draws the rope back across his chest. It may come as a surprise, but with this technique, even a small caver weighing only 125 lbs. can easily and safely stop the fall of a climber weighing 200 or more lbs. with no danger to either. After the belayer has stopped the fallen caver, he slowly lowers the climber back to the bottom or to a safe ledge where he can gather his wits and start climbing again.

For an upper belay the usual position is standing with the rope running lower down around the seat. With an upper belay it's important to realize that the direction of force will come from above and the feel of the rope is a little bit different, although easily sensed in practice. When an upper belayer is anchored with a sling or rope tied to an anchor point, the belay line should run above the anchoring or tie in line.

As when rigging for a rappel, the climber tying into a belay line must stand clear of the edge and be in a perfectly secure position. When he has tied the proper bowline on a coil or snapped into a previously tied bowline knot in the belay line with the carabiner on his chest sling, he calls out, "Ready on belay." The belayer answers when he himself is ready, and not before, that "Belay is on." It's possible that several seconds or even minutes may pass before the belayer is ready to give the "Belay-is-on" signal. However, under no circumstances should the climber assume that he is being belayed until he hears the return call "Belay-is-on." Then when the climber is ready to climb, he moves over to the lip of the pit or fissure and calls "Ready to climb." The belayer answers, "Climb when ready."

It is mandatory for the tyro to learn these climbing signals at the very start of his training and to use them inflexibly in all climbing situations. The other signals needed are the words "Falling" in case the climber is actually falling, "Tension" when a caver wants tension on the rope, or "Slack" when he wants slack. Some cavers also use the phrase "Up rope" to indicate tension, although "Up rope" is more commonly used to raise the rope after the climb has been completed.

When the caver reaches the bottom, he calls, "Off belay" to tell the belayer that he can relax. However, never give the signal "Off belay" until perfectly secure and preferably sitting down. Finally the belayer calls "Belay is off." The next call is "Up rope" indicating that the caver has untied his belay knot and is ready to have the rope hauled up for the next man.

As the rope is pulled up it is placed carefully behind the belayer, although not necessarily in a perfect coil arrangement. But the line must be free enough so that in case of a fall it will pay out with no kinks or obstructions which could cause a sudden jerk and hurt the falling climber.

In general, belaying in caving is somewhat easier than in mountaineering because if a fall is encountered, it is usually quite short, amounting only to the length of the slack rope above the climber. This means that if a relatively tight belay is kept on the caver, he will probably fall only three or four feet. Even with a loose belay, the fall will only be 10 or 12 feet at the most. In mountaineering, on the other hand, a climber may fall 30 to 50 feet or more, causing a much greater shock on the belayer and on the line.

Anchors

Safe rigging of lines for belaying, rappelling, prusiking, and ladders depends on secure anchors. Suitable anchors outside a cave when rigging an entrance drop can often be found on the large rocks or trees nearby. Inside a cave, large breakdown boulders, limestone outcrops, and some massive formations can also be used. However, be wary of formations for conservation and safety reasons. They often have a concealed structure that is quite irregular and makes them unexpectedly breakable.

For maximum safety, *separate* anchors are mandatory for each belay line, rappel line, or ladder. Often a double anchor is better for a line or ladder if there is any doubt about the security of an anchor. In any kind of anchor use the maximum cross section, particularly with limestone outcroppings. Separate slings are often used around the anchor itself to save some length on the main climbing rope or to position a rope or ladder more properly. Remember, as mentioned earlier, use as long a sling as possible so that the angle between the two sides of the sling is as acute as possible. (See Figure 7-4.)

When suitable natural anchors are not present, cavers usually turn to bolts or to the newer nuts and wedges to form a suitable anchor. It's well to keep in mind that the recommended anchor strength for stopping a belayed fall is 2,000 lbs., that a rappel or prusik anchor needs 1,000 lbs., and a direct aid device like a ladder needs 500 lbs.

FREE CLIMBING

Free or balance climbing (as opposed to tension climbing—chimneying and scrambling—see chapter 6) isn't required all that often in caving (probably only 5% to 10% of the time), but it is still necessary in order to fully explore many caves. A novice should learn the fundamentals in practice climbs above ground, before trying his prowess underground. In caving, climbs are usually made up of short—that is, 10 to 25 feet—pitches rather than the longer pitches found in mountain climbing. This makes cave climbing more relaxing and somewhat less difficult. The principles of good climbing are balance and pace, resulting in a smooth, almost measured movement. Proper balance comes from having the back of the head directly above the ankles whenever possible. When necessary to bend, keep the back very straight and the head up, bending only at the ankles, knees, or hips. When bending down to look for a hold, bend the head, but not the rest of the body. If properly standing and balanced, a caver's hands don't actually grip the handholds, rather he uses these for balance.

Good free climbing involves having three points of support at all times, then moving to a new hold while maintaining the first three. In general, it's best to use holds that are close together rather than having to stretch out in a spread-eagle stance. The reason for this is that measured movement, not awkward lunges or jumps, is the ideal. Speed is never the object in underground climbing because the light doesn't fade at sunset as in rock climbing.

Before starting up a climb, plan the route ahead, but allow for an alternate route if difficulties are encountered. Select foot and hand holds carefully, but test them before putting full weight on them. Surprisingly enough, foot holds so small as to be barely noticeable are used every day by experienced climbers. When no obvious ledge or crack presents itself for a hand or foot hold, a hold can sometimes be created by jamming the hand or foot into a crack. Be sure the foot can be pulled out, however, before putting full weight on it, just in case.

For foot holds the side of the foot is better than the toe because the toe holds put greater strain on the leg muscles. When first placing the foot onto a foot hold, flex the ankle a little bit a few times to be sure and get as much boot as possible on the hold. This helps one sense the hold with the sole of the boot, but don't shuffle around once the foot is positioned or traction may be lost. Keep the knees bent slightly for balance rather than locking them together.

Generally, climbing up is easier than climbing down because the climber can see the holds better. When descending, try to face outward when it's relatively easy going, move sideways when it's a little harder, and face in (as if climbing upward) when it's very difficult. But whenever possible, face outward when going down because the holds can be seen more easily.

When climbing down, always use a belay if there is real danger. Never be afraid to ask for a belay if feeling uncertain about the safety of the climb. On a tricky descent, the caver should ask for tension from the belayer so that he can lean out and look over his shoulder to find the next hold. In the worst case, the belayer can hold the climber securely allowing him to probe around with his foot until he finds the next hold. Be sure the belayer understands what is happening however, before doing this.

As with tight crawls and chimneys, the experienced cavers should talk a new caver down a slope, explaining where the holds are and where to make his next move. Others in the party can also help out by shining their lights on the holds so that they can be seen more easily in the darkness.

In caving, when a climb down is particularly long or arduous, a line is usually rigged and the descent is made by rappelling. Similarly, if when coming out a particularly long climb is necessary, it will often be rigged for a ladder climb or for a prusik. Or, if a climb up is really too difficult, an alternate route is sought (and usually found) that avoids it entirely.

A final note of caution. One should never climb up a slope or pitch that he won't be able to climb back down easily. Sometimes it will run into a dead end and it will be necessary to climb back down. If it's not too long a pitch, it can sometimes be rigged for a hasty rappel, using the universal slings carried by two or three cavers.

WET SUIT CAVING AND CAVE DIVING

Neither cave diving nor wet suit caving are for beginners. In particular, cave diving has claimed dozens of fatalities in North America. It combines the worst dangers of scuba diving and caving. No beginner caver should ever consider diving in a cave, even if he is a fully experienced scuba diver, without first joining

Figure 8-13. Climbing Down. When climbing easy pitches, face outward. On harder descents, turn sideways and use opposite wall if close enough to reach with hands, feet, or back. On steep ones, face inward as in climbing up, using three points of contact and classic hand and foot holds.

a cave diving group to gain the proper experience and training.

Wet suit caving is not quite in the same danger class as scuba diving in a cave, but is has its hazards, too. A maneuver called "duckunder" or "siphoning," relatively more common in British caves, is done rarely on this side of the Atlantic. The siphon or duckunder is a place where the ceiling of a passage dips down into a pool or stream so that there is no breathing space left. This usually occurs after the group has been walking chest deep or crawling on hands and knees in freezing cave water.

If the group knows that a duckunder is only a few feet long and they have negotiated it before successfully, they may try it. If it's a new one, they may decide to tie a belay line on the first man through, to serve as a handline for the others. However, as a beginner, one should never be afraid to say that his limitations stop here. Never attempt any maneuver in a cave, especially this one, until fully confident and feeling it is perfectly safe. Duckunders are for very experienced people who have done them a number of times before and are properly dressed with wet suits and waterproof equipment containers.

ADVANCED CAVING QUALIFICATIONS

After a caver has had 30 plus hours of experience on cave trips with an organized group and has met the suggested qualifications for beginners (see Chapter 6), he can consider himself in the advanced category if he can meet these qualifications:

1. Demonstrate the proper verbal signs and use them in each test.

2. Belaying—Find and rig a satisfactory anchor for a belay line in a cave.

3. Belay a 200 lb. caver in a cave from both the upper and lower belay positions, successfully holding an unannounced fall from both positions.

4. Ladder Climbing—Find and rig a satisfactory anchor for a cable ladder.

5. Climb *down* and *up* at least a 30 foot cable ladder with a proper dynamic belay from a separate belayer. Also, climb up and down a 30 foot ladder using a self belay on a fixed line with a prusik knot or mechanical ascender.

6. Rappelling—Find and rig a satisfactory rappel anchor in a cave.

7. Rappel in three different types of descents in a cave: a tight fissure or pit, a medium width fissure or pit, and a wide fissure or pit with free fall. Each drop must be at least 40 to 50 feet.

8. Prusiking—Find and rig a satisfactory anchor in a cave for a prusik.

9. Prusik up the three separate types of drops required for the rappel test (in the same or a different cave).

10. Free Climbing—Climb both up and down while on belay, a 20 to 30 foot pit or wall that is too wide for chimneying, using hand and footholds.

9

Going Caving

What's it like going on a first caving trip?

First of all, take it as a popular misconception that caves are all like Mammoth Cave or Carlsbad Caverns with huge rooms and tunnels large enough for a subway train to speed through. The fact is, most caves are small, wet and short. They have walking passages, crawlways, crevices, narrow fissures, canyons, arches, bridges, waterfalls, pits, and domes. A caver must be prepared to walk on floors covered with jagged rocks and stones, crawl through mud, climb on sharp rock, slosh through running streams, scramble up and down slippery slopes, or any combination of these in quick succession. With that in mind, let's see what a typical trip is like in the National Speleological Society.

THE ORGANIZED CAVING TRIP

Most caving trips originate with some kind of discussion at a grotto or club meeting, during which the trips for the coming week or month are planned. This planning can be as informal as a grotto chairman or trip coordinator asking "Anyone want to go to Crystal Caverns or Man Trap Cave this month?" Or it can devolve into a serious blackboard session with the pros and cons of each trip listed and argued thoroughly before a firm schedule is arrived at.

From these discussions, a novice can get an idea of what each cave is like, whether newcomers to caving are welcome on a given trip, and what equipment may be needed. Remember, that the trip leader is the one to say who can come along and what qualifications will be needed by each caver.

Usually, after the schedule is set, specific details are ironed out like "Who's going to drive and who has room for riders?" "Who will bring the group equipment like ladders, ropes and extra carbide?" "Who's planning to camp and who's coming over first thing in the morning?"

Camping Out

Camping is becoming increasingly common among cavers, because very few grottos seem to be lucky enough to have good caves within a 10 to 100 mile radius. On the contrary, a surprising number of groups routinely drive anywhere from four to eight hours each way on a typical caving weekend.

For this reason, many trips leave the night before, especially for Saturday or full weekend trips. Friday night, then, is often a hectic time right after work or classes let out. The group usually gathers at one or more rendezvous points, and gets off on the trip about 7:00 o'clock, planning to arrive sometime before midnight.

Usually, the grotto will have a regular camping site near the cave at a state park or similar campground, or there may be a club house maintained by one or more grottos. All that most cavers need in a campsite is sleeping bag space and bathroom facilities (sometimes these can be a little on the primitive side, sometimes they are very uptown). Lately, camper type vehicles, either the pickup or Volkswagen van type, have become popular among those who go caving a lot and prefer to bring more than the barest creature comforts along with them.

For the group that leaves the following morning, or for Sunday trips, the starting time is usually the crack of dawn so they can drive while the traffic is still light and arrive in good time for the trip. Usually, they don't have any trouble arriving on time, because cave trips don't get moving all that early on Saturday or Sunday morning. At least they don't seem to get moving as early as the more super organized types would like. By and large, most trips are scheduled to leave the camp site (or similar meeting place for those coming over in the morning) at about 10 a.m. But, as a matter of practice, they often don't get underway until nearly 11. This means that the trip may not get underground until about noon, depending on how much of a drive or hike it is to the entrance.

With this in mind, all cavers including those just starting out should plan their meal times accordingly. The important thing is to have a full meal within two hours of actually going caving or setting off on any kind of strenuous hike to the entrance. Eating a meal at the last minute doesn't leave time for digestion. The meal should be made up of familiar foods easily digested. Just before a cave trip is no time to experiment with exotic cuisine.

Length of Trips

A typical cave trip these days in any kind of respectable system always seems to last a minimum of six to ten hours. However, trips involving new cavers should be limited to two or three hours on the first encounter. After a few trips, a caving aspirant can probably handle a six or seven hour trip without any trouble, but be careful about expecting a raw neophyte to have the kind of experience and endurance needed for those 16 to 20 hour marathons that many charging cavers do routinely.

Suiting Up

At the cave entrance, most cavers rest for a few minutes to take the opportunity for a final check on their equipment. It's faster if everyone has restocked his pack at home before leaving on the trip. If the hike to the entrance is long or steep or the weather is on its usual warmish side, many cavers carry their hats and coveralls in a ruck sack or back pack, and can be seen struggling into their muddy caving suits at the entrance rather than back at the car.

Finally, everyone dons his hard hat, and amidst the popping of lighted carbide lamps or the curses of their owners when they fail to light (which are often echoed by the electric crew as they discover a loose cable or a burned out bulb), the trip begins slowly to get underway, as one by one each caver enters the cool darkness of the entrance.

CAVE OWNER RELATIONS

With that overview of a typical trip in mind, it would be well to review some of the steps taken before and during a cave trip to see what makes it successful. First of all, there is the delicate subject of cave owner relations.

Private Cave Owners

A visit to a cave always begins with proper land owner relations. Never enter a cave without obtaining the owner's permission. If he's not at home and is known to be friendly to cavers, it *may* be alright to leave him a note, but if in doubt wait for him or check back later. In either case, tell him how long the group plans to be in the cave, adding a few hours in case of delays. When coming out, be sure to check with him, again in person or by note. If for any reason he says his cave is closed, say "thank you" and leave politely. He may have been burnt by some other thoughtless visitors, and cavers have been known to end up with buckshot in their britches for being too persistent.

Cars should be parked where he suggests, no matter if it means a somewhat longer hike. Most important—leave open gates open and closed gates closed. He has a reason for their being the way they are and won't thank anyone if he has to go out and open the gate for the cows to come home or finds his garden being eaten up by pigs or goats.

As a courtesy, he should be told what the group has found in his cave. With any owner the grotto comes to know well, send a photo of the cave or even of the owner's house or barn. A copy of the cave map the club has just completed is also a nice gift, as is a Christmas card each year thanking him for his past courtesy.

Sometimes a reluctant owner can be convinced to let cavers see his cave if they prepare and sign a written waiver holding him blameless in case of injury or accident. The NSS Legal Committee and many individual grottos have sample release forms available for this purpose.

Caves on Public Land

Caves on state or federal land, as in the western United States and Canada, are not exempt from the rule about getting permission to enter. However, it's sometimes a problem finding who to ask. The NSS grotto or local caving group usually has this kind of information and has set up a procedure with the right agency. In some cases, it's necessary to write ahead of time (two or three weeks), telling the ranger or supervisor what group it is and what their qualifications are for going in. Most public officials responsible for outdoor recreation are helpful, so be sure to follow their requirements for signing waivers, letting them know when the group comes out, and returning the keys, if the cave is gated.

If after a sincere try, the grotto can't find any agency that has jurisdiction over a cave, it may be alright to visit it. However, it's often a disappointment to find that these caves on remote unattended public land are heavily vandalized for the very reason that there is no nearby control over entry.

The public agencies most often responsible for caves are the National Park Service, National Forest Service, Bureau of Land Management, and the recreational, conservation and wildlife or natural resource departments of the various states.

Cave Gates

Although many cavers feel that cave gates are a waste of time and money to install unless a caretaker or owner lives nearby (in which case a gate probably isn't necessary), the fact is, many caves are gated. Basically, the reason for a gate is to keep people (and cavers) out. The cave may be dangerous, it may be easily damaged by traffic, or the owner may be worried about liability if someone is injured in the cave. In any case, never force or break open a cave gate. It's illegal, it will make the owner very unforgiving, and it's a bad conservation practice.

TRIP ARRANGEMENTS

Assuming that the group has gotten permission to enter the cave, the next consideration is to get there comfortably and safely, since it is often stated as a truism that more cavers have been killed driving on the highway enroute to the caves than in caves themselves. Unfortunately, this is probably true, since some cavers have a tendency to leave much too late on Friday and drive all night to the cave. Then after a few hours sleep, they cave hard Saturday and part of Sunday and depart for home late on Sunday night, arriving early Monday morning. From a safety standpoint, this is obviously not a very sound practice, either for caving or travelling.

The question of which car or truck makes the best caving vehicle can lead to instant arguments at any NSS gathering, some nearly as heated as the carbide/electric or prusik/ladder hassles. One faction may argue that since the majority of the trip is made on the highway for distances that average 200 to 300 miles, a regular passenger car with large trunk is the most desirable. A separate trunk, these cavers believe, is better than the storage area in a station wagon, because it keeps muddy caving gear away from the passengers.

Another group may feel just as strongly that those last few miles of the journey, which frequently call for a bone jarring trip over rough if not impassable terrain, are the most important miles of all. For them, a caving car means nothing less than a four wheel drive vehicle, either a straight conventional Jeep, a Jeep stationwagon, a Toyota Landcruiser or Landcruiser stationwagon, or the Ford and Chevrolet versions of the jeep, the Bronco and Blazer. There are others who recommend the high clearance industrial type pickups or crew cab trucks (Chevrolet Carryall or Dodge Power Wagon) which can be equipped with four wheel drive.

Cavers have been known to travel long distances to caves in vehicles as interesting as old ambulances and hearses. These make remarkably good vehicles for camping, by the way, since there is an unusually large amount of room to stretch out. They also ride very smoothly and have plenty of storage space.

A caver who rides with a group is usually asked to share in the expenses. In many NSS grottos, this is usually computed at one cent per passenger per mile, as a matter of convenience and simplicity. This works out to be $1.00 per hundred miles. The new caver should be prepared to pay this or some other token amount if he is a rider on a trip.

ENTERING THE CAVE

Here is a short checklist that the trip leader or coordinator should run through mentally just before entering a cave. Always leave word with the owner or caretaker of the cave, or with someone back home as to the trip plans including when the group is expected to come out of the cave and what parts will be explored. Check the weather conditions for rain or threatening skies. If entering a new system, leave someone at the entrance for safety's sake. Be sure that everyone in the group is capable of handling the difficulties expected in the cave to be entered. Be sure each caver has eaten a meal in the last two hours before going underground. If not, stop now and eat some high energy food, so that everyone will have their full strength and not be distracted by hunger during the trip.

The ideal size for a caving group is a matter of some debate but it is strongly recommended that the minimum size must be four. This allows for two cavers to go out in case of an injury, while leaving one person to stay with the injured caver. The reason that two must go for help should be obvious—never cave alone. The very first rule of caving safety must especially apply in this situation. Furthermore, it is far better for two people to describe the condition of an injured person and the problems that might be encountered by a rescue team than just one.

Order of Movement

Normally, movement in a cave is in single file, with the leader or most experienced person going first, the novices or most inexperienced cavers in the middle, and the second-most experienced person at the end. A single file is most efficient because this way it is easier to pass messages back and forth from one caver to the next and maintain close contact. Single file causes less damage, too. It is also important that cavers refrain from any unnecessary talking or yelling in order that messages can be easily heard and everyone's attention can be kept directed on careful balanced movement. On any trip with a novice, the experienced cavers invariably will help out by showing the best routes and methods of moving. It is always important to show learners that any particular move can be made quite easily when done properly. Encouragement is also helpful.

Quite often the leader of a trip made with some apprentice members will keep a kind of running commentary going describing what they're coming to, the best techniques to use, and other items of general interest. This helps to dispel the inexperienced caver's fears and allows the man in charge to get a feel for how the trip is progressing.

A newcomer to caving should keep in mind that it's his responsibility to keep his eyes open on his first few caving trips, not only to note the speleothems and keep track of the route, but to be able to see how experienced cavers walk, climb, and carry themselves in a cave. Upon entering, go slow or stop for a bit to let the eyes get used to the darkness after the bright outdoors, then adjust the lamp as necessary.

Weather Conditions

It bears repeating that experienced cavers are very cautious when going in a cave with underground streams or lakes if rain is threatening. If already in this type of cave, watch out for rising streams, changes in the amount of debris, muddier water, a higher noise level than heard previously, and increased foam at the base of waterfalls. When any of these signs appear, possible danger is indicated. If the group is away from the stream at that time, they should listen and note any apparent increases in sound or changes in air movement. Air movements sometimes change rapidly when water rises so it is important to be conscious of changing air currents and investigate immediately.

If there is danger from rising water, the best plan is to leave the cave immediately. If this isn't possible, move to the highest ground available. The decision as to whether to leave the cave or seek high ground is one that the leader of the trip or the most experienced caver should make after discussion with the other members of the trip. Don't forget that a rope or ladder previously rigged in a waterfall may have suddenly become far more dangerous because of an increase in the force of the water. It might be enough to actually knock a man off a ladder.

EXPLORING

Known Caves

Before setting out on a trip to a well-known cave, check with the other cavers in the group to find out what equipment to take, what clothes to wear, and what to expect on the trip. Often one can save a great deal of time and effort through a little inquiry before setting off. If a map is available of the cave, this is certainly worth studying, although relating a given passage underground to a line on a map can be quite difficult. However, studying the map can yield a good idea of the main structure of the cave as well as the different levels and main routes. With map in hand, a person who knows the cave can usually explain to the uninitiated most of the hazards quite clearly.

By far the best way to find a known cave, new or old, is to go with someone who has been there before and knows where the entrance is. Also, cavers are increasingly reluctant to give directions to newcomers unless they themselves go along on the trip to be sure that the proper safety and conservation practices are followed.

New Caves

Discovering a new cave or a new section in a known cave is certainly one of the thrills of caving and speleology. However, it also carries certain responsibilities. First of all, if the discovering group is really the first one in, take care not to disturb anything in the way of formations, cave life, or even debris like wood and stream stones. Avoid especially beautiful sections if they cannot be entered without destroying some of the speleothems. The rule is, photograph; don't collect. If anything is found of special interest, it should be left in place and a scientist told about it.

Be especially careful in selecting a route in a new cave because the chances are that others will follow the same footprints. Always choose the route that will damage the cave the least.

From a safety standpoint, remember also that a really new cave may have loose stones, false floors or unstable ceilings that can be quite dangerous. Proceed quietly, cautiously, and slowly. Be especially wary of streams and pools because these are often far deeper than they look owing to the poor light and clarity of cave water. As in any kind of caving, the entire group should stay together with no one rushing off haphazardly. It's best to wait and do it right. Remember it is not necessary to explore the whole cave the first time in!

When confronted with a series of confusing leads, particularly in a three dimensional maze, leave a person or candle at the exits of the passages traversed so they can be easily identified when returning. Also, no caver, especially the new one, should ever leave his pack when checking a side passage. One can never tell when a light will fail and spares will be needed from the pack.

ROUTE FINDING UNDERGROUND

Surprising as it may seem, very few people ever get lost in a cave although every caver gets momentarily confused at one time or another. In the early days of caving, many cavers (including members of the National Speleological Society) used to carry string into a cave as Tom Sawyer did in his well-known underground adventures. However, it was soon discovered that a cave small enough for an average size ball of string didn't need string anyway, and a really large cave required far more string than could be easily carried, even by a group.

The trick in finding one's way underground is to turn around regularly in each passage and study it carefully, especially at junction rooms where several passages take off. Pay close attention to the passage ahead also, looking for distinctive formations, speleothems, flat ceilings or floors, or other distinctive features that can be identified when finding the way back.

Watch for obvious signs of traffic by cavers in each passage. These may be footprints, survey markers, "out" arrows, Scotchlite markers, and other signs of human presence (including, regrettably, trash). But be careful that these signs don't lead into new or difficult areas off the main route rather than leading the way out of the cave. "Out" arrows, by the way, smoked on the wall with a carbide lamp (although not a recommended practice anymore), are still seen in older passages and *normally* point out of the cave—never in. When coming to a completely new area, the trip leader or most experienced caver plus one other caver should always check ahead by themselves for the best route and the correct passage and then return to the main group. It is always very tiring and disheartening to push into a wrong passage that pinches out in 20 or 30 yards and then have to backtrack.

The correct route in a cave is seldom the most obvious one, although it does happen that the main passage in some caves is quite easy to follow. However, in others a small hole above or to the side of what seems like the main route often turns out to be the right one. Sometimes, studying a map of the cave carefully before coming in and making notes of distinctive areas and pits, can be a real help in finding the route. However, in many caves a map does not begin to make real sense until after one has made one or two trips through so the caver can relate where he has been to the passages shown on the map.

As a training measure, many groups in easier caves will explain to their novitiates that they should pay close attention to the route on the way in because they may be asked to lead the party out if they can. As long as this is done in a good natured way without any threats or implied punishment, it can be an excellent training exercise because it helps to teach new cavers how to pay close attention as they move through the cave.

TRANSPORTING GEAR

Helping others move gear in tight passages or vertical pitches is part of the teamwork of caving. Quite often this can be done by just handing a pack from one person to the next, but if the slope or pit is too deep, then the universal caver's sling can be untied and made into a single length with a figure eight loop to tie in gear for lowering. When hauling gear, remember that a string of small bundles is better than one large one because they don't hang up as easily. In the case of very large equipment, such as might be required for a cave movie or cave dive, it is recommended that a team be formed, made up of at least three experienced cavers: one at the top to lower the gear, one tied in at some point or ledge on the way down, and one at the bottom. If the descent is really tricky with many obstructions and tight spots, it may be necessary for the middle man or even a fourth caver to follow the load down, freeing it as necessary from time to time with well placed kicks and shoves.

DISCOVERING NEW CAVES

Caves occur in limestone, gypsum, or lava, although the largest number by far were created by water dissolving out limestone. Therefore, if a caver lives in a limestone area of the country where there are known caves, either commercial or wild, the chances are there are other caves nearby. Limestone or marble tends to be deposited either in broad beds covering fairly large areas or in discrete but sizable pods.

A good way to seek out caves is to ask commercial cave owners or the guides at a commercial cave if they know of any wild caves nearby. The chances are they do; they may even have explored them themselves. Check also with local farmers, hunters, miners, fishermen, hikers, power company and telephone company linemen and some of the oldtimers of the area. Frequently the oldest inhabitant will have heard of some caves in his youth and can suggest where to search for them. Try also at the local library, historical society, newspaper editors, or county recorders. With their help, read up on local history, diaries, old newspapers and county geographical or geological surveys for references to caves in the area.

Entrances

Once an area is isolated as having cave possibilities, get out in the countryside and look for outcroppings of limestone, particularly for any that have air blowing between cracks or out of small openings. Air coming from caves is sometimes easier to see in the winter, because a cave is normally warmer (45° to 55°) than the surrounding air. There may even be melted snow at cave entrances. Watch for sinkholes and streams disappearing underground, and for springs as where streams appear to come back out after travelling underground. These are indicators of possible cave entrances somewhere nearby.

Entrances are usually at the bottom of a sinkhole, near or at the bottom of a valley or gully, or a hole in the side of a valley or canyon. Some caves have been discovered by following animals who have suddenly disappeared underground. The literature of caving is full of examples of a farmer, a group of boys, or a hunter sitting on a ledge near a limestone outcrop and seeing a small animal like a rabbit or a woodchuck suddenly disappear. Careful examination of the area usually then revealed a small hole or crack, which was opened by moving a rock and digging until the entrance of a cave was uncovered. The beautiful caves of Lascaux in France, with their treasure of prehistoric art, were discovered this way. Fortunately, the four boys who stumbled on the entrance while trying to find their lost dog had learned about ancient man in school, so they immediately told their teacher about the find. Within but a short time scientists and art experts from all over France were studying these masterpieces that are among the most beautiful and best preserved (because of the sealed entrance) of any prehistoric art.

Discovering new caves requires a certain amount of detective work (and luck), but almost anybody can do it successfully. If a caver enjoys looking through old historical records, doubtless many caves known in earlier times can be found. This was proven in the late 1940's when the Stanford Grotto of the National Speleological Society was able to relocate countless caves by careful study of old newspaper records, county registers, and similar reports from the early gold rush days of mining in California. It seems that the 49'ers investigated any number of caves hoping to find that elusive vein of gold. There is no record of any significant gold find ever having been made in a cave, but the old time miners checked them out anyway, ever hopeful of finding that biggest strike of all. Tales of the underground splendors reached newspaper editors who were always looking for a good story and were not above embellishing the bare facts to make it more interesting.

One distinctive sign found in many caving areas is red silt or clay. It seems that many cave-bearing rocks of limestone also have some iron minerals intermingled with them which cause a very distinctive reddish color in the mud and clay. There are some cavers who believe all caving areas have red clay. While this is not true, in colder areas, it often happens that where there is red soil one will probably find caves. Another good indication of caving country is an area which is known to have a great deal of rainfall but where there are only one or two major rivers on the surface. Smaller brooks and streams are strangely absent. This generally means that surface water quickly goes underground and flows through channels in limestone. This kind of terrain is called "karst" and it is usually marked by a number of small or large sinkholes. These may not be so obvious from the surface, but when studied from above as from an airplane or high hill, they stand out quite clearly. Should one come across a sinkhole that has a small stream pouring into it with no obvious exit, there is more than likely a cave and a cave entrance at the bottom of that sink.

DISCOVERING NEW AREAS OF KNOWN CAVES

Even if a caving group is unable to discover any new caves, there's no reason to be discouraged. Many very large cave systems have been discovered, one piece at a time. Mammoth Cave in Kentucky, for example, had been commercialized for many years before one of the guides on his day off pushed into a vast new area. This was later commercialized and added to the standard tour.

When inside a cave, look for areas that are obviously filled with dirt or large breakdown blocks, which with careful excavation can sometimes be penetrated to find more cave. When a passage is filled with mud or breakdown, it usually results from flooding or a collapse in the roof which blocks that point. However, the passage often continues on the other side.

If a group finds some possible cave leads, they should discuss them with the land owner and get his permission to proceed farther. If any surface digging is required, be sure to fence the dig off in order to prevent harm to farm animals. Many caves have been discovered by digging, but the procedure has its own special hazards. If the dig is going to go any distance at all, special care must be taken to shore up the sides of the excavation to prevent collapse. Also, if any blasting must be done, it is obvious that the services of a licensed professional are mandatory.

When seeking new leads in a known cave, a surface survey can sometimes reveal the most likely direction of a new area. Quite often promising leads, when correlated with surface features, will be found to end at a sharp cliff. Or, a resurgence of water may be seen indicating that no more cave may be found in that particular direction. This is where a good map of the cave coordinated to an accurate surface survey is of great value.

Sometimes, a well-known surface stream one day suddenly disappears into the ground into what is called a swallow hole. The actual size of a swallow hole is not too good an indicator of the possible size of the cave below. It may turn out that a large swallow hole will lead into a very small passage and that's why there's so much water up above. Or a small swallow hole may be small because it is pouring water very rapidly into a large passage hole. Be very careful when pushing a lead made by a swallow hole. Many times the rocks are rather precariously perched on the sides of such a hole and it may take no more than the pressure from a caver's boot to cause a cave in.

10

About Caving Specialties, Accidents and Hodags

Not all of a caver's spare time is spent crawling about, exploring dark underground passages or figuring out how to get around, over or somehow past the seemingly impossible barriers he sometimes encounters along his route. In fact, a great deal of his time may be spent in adding to the fund of knowledge available to students and scientists important specific data about individual caves such as precise measurements, additional photographs and notes on unusual items noted.

CAVE MAPPING

Many cavers and caving clubs make a regular practice of surveying caves and preparing cave maps. A cave map is useful because it can tell a great deal about the cave and if correlated with a surface survey, can provide a good clue as to where to look for additional passage in known caves.

Cave maps are basically a plan of the cave, that is, a view of the cave passages and rooms as they would appear if one was looking directly down through the earth to the passages below. In this respect the cave map is like a map of city streets. In making a cave map it is also helpful to prepare cross sections of various points in the cave to show the vertical extent of any pits or differing levels. In fact, if the cave has much vertical extent, a vertical cross section or profile through the major portion of the cave is often prepared at some time later.

Cave Surveying

Surveying a cave involves making a single-line plot of the main passages, plus any important secondary passages as may be desired. Basically, surveying requires the caver to measure the azimuth or direction of the passage plus its slope either up or down. Several compasses are available that make both of these measurements (the compass needle for azimuth, a clinometer

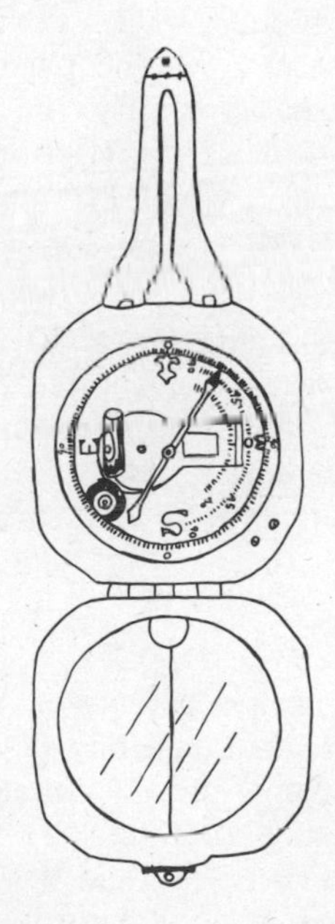

Figure 10-1. The Brunton Compass. A rugged cave surveying instrument; measures azimuth with a compass and vertical angles with a clinometer. Circular bullseye level at lower left is used to level compass for azimuth readings. Oblong level at center is for angles, as read from scale at right.

for slope). Perhaps the most common in use by experienced mappers is the Brunton compass which is specifically designed for surveying. It can be either hand held or used with a tripod. In cave use, the hand held operation is nearly universal. The Brunton Pocket Transit model sells for about $50 to $60. There are also available several compasses in a lesser price range, notably those made by Silva and Suunto in a variety of models. The liquid filled versions are preferred by many because their dials or needles tend to stabilize more quickly, but some cavers prefer those with merely a finger operated brake since once azimuth has been established and the brake put on, the compass can be lifted to any position more convenient for reading the dial, an aid to those wearing spectacles, for instance.

Some of the Silva and Suunto's also have a clinometer like the Brunton for measuring vertical angles. A homemade clinometer can be made (as first suggested by Alan Budreau of the Boston Grotto) from a protractor, spirit level, wooden ruler, two paper clips, and an inspection mirror (like a dentist uses). These are assembled as shown in Figure 10-2. For sighting, two quarter inch straight pieces of metal from the paper clips are epoxied to the left side of the protractor, so that they stand straight out at right angles. In use, sight along the clips from one station to the next, rotating the protractor while keeping the bubble between the lines in the spirit level. Read the angle from where the bottom of the ruler intersects the protractor scale.

To take distance measurements, a 50 or 100 foot steel tape is used. Some cavers have used a knotted rope or even television lead-in cable for distance measurements, but tape is much preferred. In a pinch, if no better means are available, pacing can be used, since any measurement is better than none.

Taking Readings

Azimuth and angle readings are taken from one station to the next by at least a two caver team consisting of a surveyor and a surveyor's helper. This second station should be at a point permitting the longest sighting that can be taken along the passage. The second man directs his light back toward the first station so the instrument man can sight on it. A three man or even a four man team is better, so that one man does nothing but record the readings, one man operates the instrument, and two men handle the tape. When surveying, woolen clothing, especially

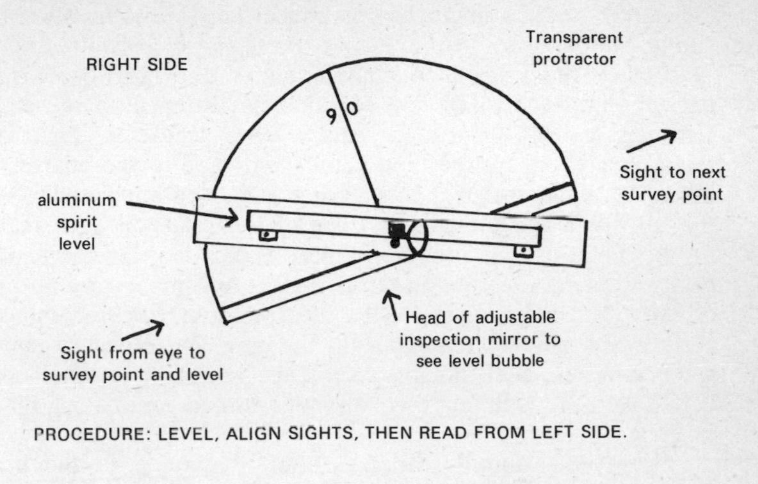

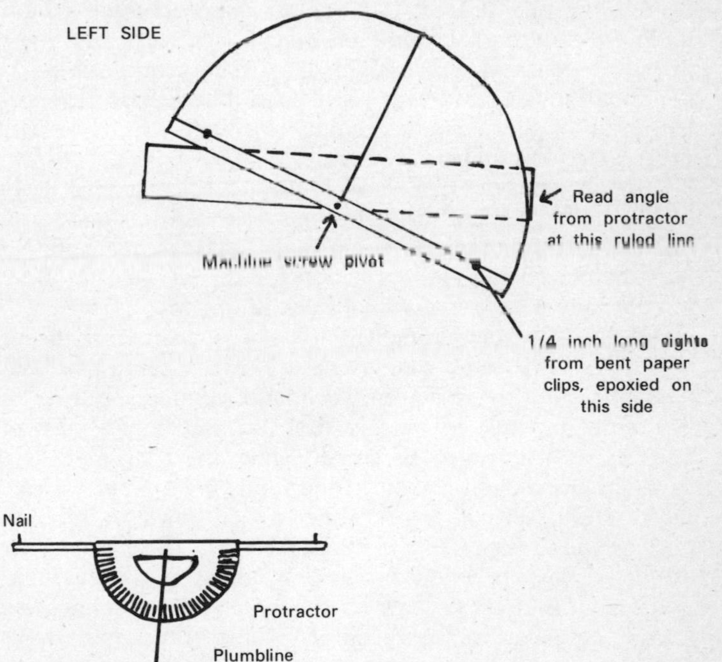

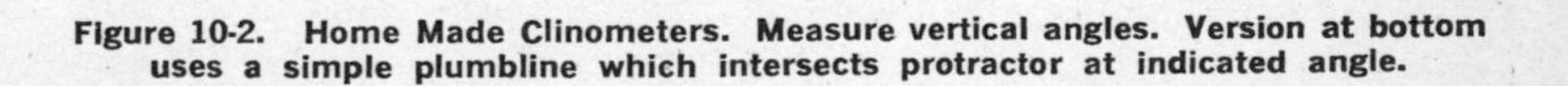
Figure 10-2. Home Made Clinometers. Measure vertical angles. Version at bottom uses a simple plumbline which intersects protractor at indicated angle.

long underwear, is mandatory, because the process involves much sitting and waiting.

Follow the instrument manufacturers directions when taking the measurements. It's best to take one's time and recheck the sightings at least once to be sure they're accurate. To measure the distance, stretch the tape tightly and read to the nearest foot. For most cave surveys, considering the conditions under which they're made and the eventual use of the map, distance readings to the nearest foot are more than adequate. However, if the group wants dimensions in terms of feet and inches, by all means do so. Readings only to the nearest foot are recommended merely because they're simpler in the cave and easier to interpret when one gets back home. Any inaccuracies in distance are always averaged out, anyway, and distributed among all the stations.

However, azimuth readings should be made to the nearest degree (or half degree, if possible), since correct azimuths are far more important in the final map. If there is any question on azimuth accuracy, a backsight from the second station back to the first can also be made as a cross check. Be sure to record the location the azimuth reading was taken from by citing the proper station numbers in the log.

Some compasses can be set to correct for magnetic declination, so that true compass readings can be read directly from the dial. The main thing to remember is that the area's magnetic variation must be considered and correction for this applied to all azimuths at some point before the final map is drawn.

Figure 10-3 shows how the data is entered in a surveying log book. This book can just as well be a small notebook from the dime store as an expensive bound volume, because the cave environment tends to quickly destroy almost any kind of notebook in a short time. Essentially, the surveying team is reading from the instrument station to the following station and recording the azimuth, vertical angle (either up or down as indicated by a plus or minus sign), the distance between the two stations, plus estimated passage width and ceiling height at the instrument station. On the opposite page of the notebook a rough sketch map shows the passages, the position of the stations in the passage and the rough shape of the passages. Backsights (if any) are simply shown (see Figure 10-3) as a reading from station four to two, right after the reading from station two to four. The other measurements (slope, distance, and passage dimensions) do not have to be repeated when taking a backsight.

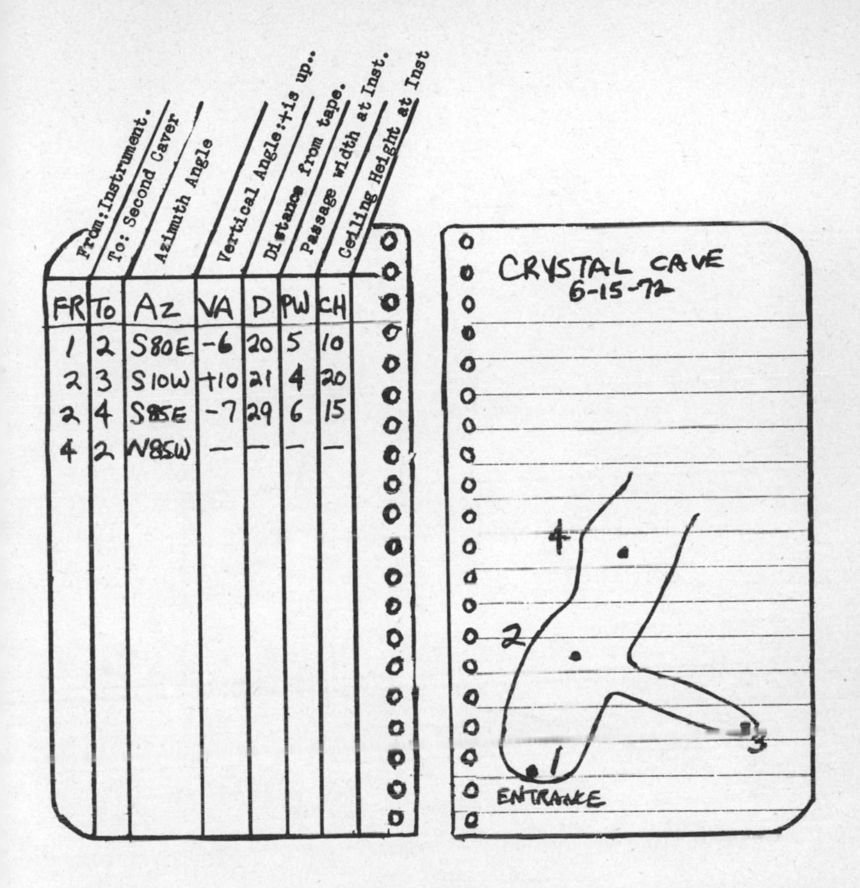

Figure 10-3. Typical Cave Surveyor's Notebook. Note that a backsight is simply shown as a sighting from station 4 to 2, the frontsight being from 2 to 4.

Drawing the Cave Map

When drawing the cave map, first choose a proper scale. Cave maps can run anywhere from an 8½ by 11 inch piece of paper to huge rolls three feet wide by 15 or 20 feet long. The amateur cave cartographer should pick a scale that will enable him to make his first sketch on a reasonable sized paper, probably 8½ by 11 inches. To plot the azimuth angles, a simple protractor and ruler can be used, although the instruments on a draftsman's table makes the job considerably simpler.

Before the distances between survey stations can be properly computed, a simple trigonometric conversion has to be made to

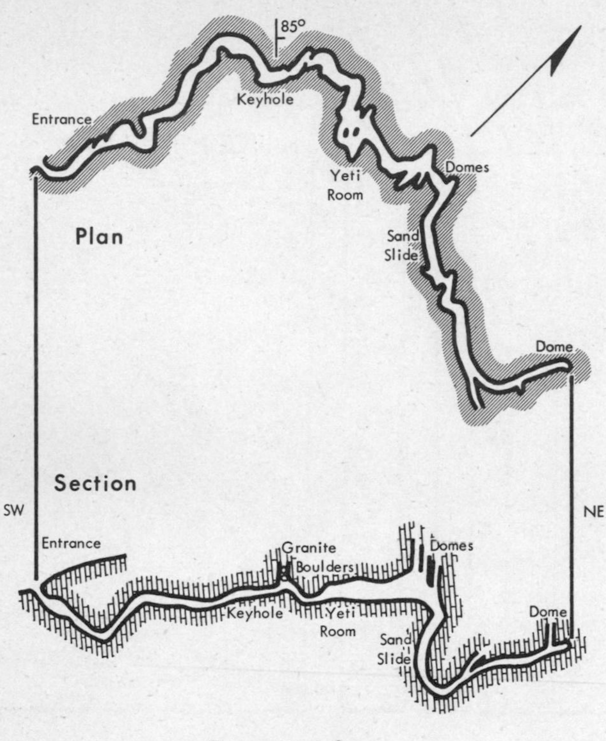

Panorama Cave

36.4°N., 118.5°W.

George W. Moore

1964

0 50 100 Feet
0 10 20 30 Meters

Figure 10-4. Completed Cave Map. Shows both plan and section. Scale is chosen to fit size of cave, size of final map, and amount of detail to be included. Scale does not specify "one inch or cm equals so many feet or meters." This allows it to be reduced or enlarged in printing without losing size relationships.

convert the slope distance, which reflects any upward or downward slant of the passage, to a true horizontal distance. This is done by multiplying the slope distance times the cosine of the vertical angle. Any standard handbook of physics or reference book with mathematical tables can provide the cosine figures for various angles as expressed in degrees.

The first step in making a map is to lay out the entire cave with a single line connecting the survey stations. Then from the data showing passage width and the rough sketch showing position of the stations in the passage, detail can be added to fill out the passage shapes. Normally have north at the top of the page and always provide a North arrow. Show only true north on the map, making any conversion necessary from magnetic north before drawing the map.

By far the most accurate map is going to be one in which a loop or circle connecting back on itself can be made. This, of course, depends on the nature of the cave, itself. However, most caves can be broken up into sections for surveying purposes and several loops made. The advantage of making a loop in cave surveying is that any errors made in angle or distance can be averaged out over the 15 or 20 stations in the loop without too significantly affecting the accuracy.

Standardized symbols for cave maps are covered in the NSS Caving Information Series No. 16, available from the NSS office. Surveying techniques (a short discussion), are contained in No. 2 of this Series.

So far, this discussion has been mainly concerned with making a relatively simple cave map such as any newcomer to caving could make with some training from a local grotto. If more detail is desired, cross section data at particular points can be assembled on a return trip and these then added to the map. If the cave has considerable vertical extent, a full length profile or vertical cross section can also be made up, but that is beyond the scope of this text. A fuller exposition of surveying and mapping can be found in the *NSS Bulletin* 9 and in *British Caving.* (See bibliography for details.)

CAVE PHOTOGRAPHY

Cave photography is far and away the most popular extra activity taken up by cavers. Choice of the correct camera is pretty much an individual matter, nevertheless a few guidelines will be offered here.

For anyone who is not an experienced photographer and who does not already own a camera, it is suggested that one of the simpler instamatic type cameras with drop in film cassettes and plug in flash cubes will be more than adequate as a starter. Even in advanced cave photography, some photographers recommend the flash cube over other sources of light because of its great convenience.

For the more advanced photographer, the most common type of camera found in caves today is undoubtedly the 35mm single-lens reflex. Since these cameras come in a mind boggling array of options and special deals, no specific brands will be recommended. (They've all been used at one time or another underground.) Just be sure to select one that is easy to operate, both in terms of the placement of the controls and ease of use with cold muddy hands in the dark. Ruggedness is also an obvious virtue. The most often used lenses will undoubtedly be the so-called normal or 50mm lens, but the wide-angle lens of anywhere from 19 to 28mm also finds use in cave photography, as do close-up accessory lenses.

Almost universally, the cave photographer uses his camera hand-held to shoot color slides for projection at meetings and conventions. A few more serious photographers may also use tripods and multiple-flash set ups to shoot negative color for color prints or black and white negatives for enlargements. These are almost always destined for submission in the annual NSS photo salon.

Camera Packs and Cases

Minimal protection in easy horizontal caves can be provided photography gear by a canvas cave pack with the camera installed in its own leather case inside a plastic bag. However, for serious cave photography, a surplus army ammunition can padded with foam rubber is the most common item seen. Some cavers even make up fitted wooden or aluminum cases for their cameras. One particularly versatile set up (thanks to Charlie Larson of the Oregon Grotto) is a platform that has mounting facilities for both a black and white and color camera, which in turn mounts on a tripod. The wooden platform is then covered entirely by a wooden case which has a handle for carrying purposes.

One of the big problems in cave photography is keeping the hands clean. This makes gloves absolutely mandatory. Condensed breath showing up in pictures, fogging of the lens, and dust are also problems, but these can be taken care of if the photographer is conscious of them while preparing to shoot.

Lighting for Cave Photography

Total darkness in a cave actually simplifies photography to some extent rather than making it the other way around. For example, the total darkness allows several photographers to ride the same flash bulb. With this technique, each of them opens his shutter on the count of one, then one of them fires off a flashbulb on the count of two, and on three all shutters are closed. Similarly, lighting a large room in a cave, although always somewhat experimental, is actually easier because of the total darkness. The technique is to open the lens of the camera, then shoot off a series of flash bulbs as the photographer or his assistant moves from point to point in the cave shooting off one bulb at each location. The camera, of course, must be kept on a steady rest. To avoid undesirable ghost-like effects, however, only one bulb should illuminate any part of the scene where a human figure will appear. For shooting multiple flashes, a cable release with a locking device to keep the shutter open in the Bulb (B) setting is essential unless the camera has a Time (T) setting—not too common a feature in modern 35mm cameras.

Electronic flash, the flash bulb, or flash cube are the universal lighting techniques for cave photography. In earlier times, flash powder was sometimes used because of its higher light output. However, it has its own special problems because it's hard to determine the proper charge and it creates smoke which must clear away before more pictures can be taken. Either the flash cube or the folding flash gun with replaceable flash bulbs are most often used. When selecting a folding flash gun, one with a tilting head with two or more settings is often desirable. Electronic flash is also widely used in cave photography because of its greater convenience, although its light output, except for the larger, more expensive units, is quite a bit lower than flash cubes or flash bulbs.

A special problem in cave photography is getting proper focus because of the low light levels. This is one of the reasons why the wide angle lens is used to a great extent since is simplifies focusing problems. The most common method with reflex cameras is to focus on a light held by the photographer's assistant. For cameras without a coupled range finder, distances can be estimated or a tape measure used if focus is critical with a given lens.

Showing Slides

Color slides are always welcome at caving meetings although a caver will be doing everyone a favor if some presorting of his slides is done before the program. Weed out all unsuccessful photographs as well as the ones of the kids in the back yard. Save these for another time. Also, remember that while the average caver wants to know what he is looking at in a picture, he also doesn't want a five minute lecture on each slide! Incidentally, the biggest single failure in cave slides, aside from poor lighting, is lack of scale. Always include something, preferably an entire person or some recognizable portion thereof such as their head or hand, or an inanimate object of known size like a carbide lamp, a small ruler, a hard hat, or second best, a flash bulb. The things most commonly photographed in caves are speleothems, but don't neglect activities by fellow cavers, like climbing, stopping for a snack midway through, or emerging from the entrance.

Other Accessories

The serious photographer uses a tripod for his cave pictures. Ideal characteristics of the tripod for cave use are obviously light weight and small size, but also keep in mind that the legs should be of a type that are not likely to jam when covered with cave mud. In general, the types that have a screw fitting which tightens onto the lower section will cause problems in cave photography. A light meter is unnecessary for cave pictures, because there only the most sensitive meter will give any reading at all. Cavers universally use the guide numbers on their flashbulbs or electronic flash, to set exposure.

CAVE ACCIDENTS AND EXPOSURE

Cave accidents, fortunately, are usually of quite a minor nature. The most common ones are sprained ankles, strained muscles, scratches and bruises, and some degree of exhaustion. Serious cave accidents, including deaths, are usually caused by falling, loose rocks, exposure, and drowning.

In general, most cave accidents involve inexperienced people. A good caver knows his own limitations and never exceeds them, realizing that putting himself in danger puts the entire group in danger. For this reason, every member of a caving party soon learns to watch himself as well as the others for signs of exhaustion and exposure.

The old caving rule is, *don't hurt yourself*. This is another way of saying that every caver is really responsible basically to himself to know his limitations and take care of himself properly underground.

Exposure or Hypothermia

Exposure is the greatest potential danger in caves. It can kill all by itself. And, as John Walker said in his excellent article on hypothermia in the January 1971 *DC Speleograph,* exposure kills just as dead as a 150 foot fall or getting run over by a truck. Shock, which is a degree of exposure, is the other major problem in cave accidents, because it invariably accompanies even minor injuries. In most cases, there is far more danger from shock and exposure than from the injury itself.

Medically, exposure or hypothermia is a failure of the body's temperature regulatory mechanism caused by prolonged activity, cold, exhaustion, and wetness on the body. It can result from a severe soaking after falling in a stream or being trapped under a waterfall. It is basically a condition where the body is losing heat faster than it can replace it. The result is a lowered body tone, followed by a coma, then death. Continued exposure is dangerous, because it builds up and is usually accompanied by a severe loss in will power and desire to resist. Doubt and worry can increase the chances of exposure, because they drain a caver's mental energy and hence his physical energy also. This is why the proper mental attitude can't be overemphasized. This is not to say that a group of cavers must march happily through the cave, each trying to make hilarious jokes. But a common feeling of good spirits will bolster the attitude of everyone in the group.

Wetness often causes exposure because it lowers the body temperature. Whenever there is a choice between a wet and dry passage, stay dry, especially on the way into the cave. When a caver gets wet, the immediate effect is cooling since the water replaces the air which provided an insulating barrier. Then, since water is a much better conductor than air, it tends to conduct body heat away from the caver much more quickly than air could do. The body responds by curbing circulation to the wet area and a state of shock can easily follow. This is why any caver involved in extensive wet caving must be properly dressed in a wet suit to retain body heat. Even in an average cave, woolen clothes, particularly woolen underwear, should be worn. As pointed out earlier, recent tests by mountaineers show that the wet cotton clothes conduct heat away from the body so fast that they are worse than having no clothes on at all. Don't take chances when a simple thing like wearing long johns could possibly be life-saving.

To some extent, food can restore body heat, particularly a warm drink like hot chocolate or tea. But it's far better to eat the proper food before getting wet in the first place. A caver must always remember that a full meal of familiar, easily digested food is essential before any cave trip. It is especially mandatory before wet caves where heat loss can be compensated for to some extent by the food in the body.

Suggested treatments for exposure which are also very good for shock are: 1) move the victim out of water. This is a must. 2) Replace wet clothing with dry to improve insulation. 3) Insulate him from the ground. 4) Move him out of air currents. 5) Keep his head warm, since the head and neck account for half of the body heat lost by radiation. 6) Try to keep him awake and warm him up with down clothes, direct body contact, carbide lamps, or whatever.

Cave Rescue

Cave rescue is an extremely specialized caving technique. Fortunately, caving accidents are relatively rare in the Americas and the need for rescue teams so far has not required an extensive network of regional teams in constant readiness for rescue work.

If a beginner is on a caving trip where a serious accident happens, he can make himself useful by keeping as cool as possible and following directions of the leader or more experienced cavers right to the letter. When an injury occurs in a cave, the reason why a minimum of four is necessary for a caving trip becomes obvious. Two people must carry the message out to get help. This is because the rule of never caving alone applies just as strongly when someone is leaving a cave to get help as it does at any other time. Furthermore, two people can much better remember all of the details of the accident than a single person can. Whenever possible and before leaving the injured person, write down the extent of the injuries, the exact location of the injury in the cave, the surface location of the cave, itself, and the number of people in the party.

The fourth person in the minimum party of four must stay with the injured person to keep his spirits up and to minister to his needs. As stated before, exposure and shock are the most serious problems. Therefore, everyone should contribute clothing to lift the injured caver off of the wet floor and to cover him up as much as possible. If necessary, those leaving the cave can contribute some of their clothing since they will be involved in moving quickly and continuously which should be strenuous enough to keep them warm on the way out. Sometimes a low wall can be built up of ropes, clothing or packs to stop drafts from reaching the casualty. Assuming that he is conscious, a warm drink of chocolate or tea will be quite a spirit lifter and will help the others, too.

A fundamental decision must be made by the leader and a consensus of those in the party as to whether to move the injured person or whether to have him remain where he is in the cave. This is an important question, since in some cases the danger of exposure in a wet cave where no extra clothing or downfilled clothes are available may be more serious than moving the person out.

One benefit of moving an accident victim out of the cave whenever possible or at least towards its entrance is that it tends to warm him up and gets him closer to rescuers who may already be on the way. It also gives the others in the party something to do. However, one must go slow to be sure to avoid any further injury which could increase the shock.

The actual techniques of rescue involve the use of litters, ropes, pulleys, and a great deal of strength provided by as large a number of experienced cavers as can practically help out. No attempt will be made here to describe the special techniques required. The detailed information can be found in *A Manual of Caving Techniques, American Caving Illustrated,* and *Caving and Potholing.* (See bibliography.)

Public Relations

Be especially careful of involving local news media when an accident occurs in a cave. The sensationalism of injury to an underground explorer is of great interest to the average newspaperman, so much so, that distortion in reporting cave accidents is rampant. For this reason, make contact only with the local civil defense, police, fire departments, or other authorized rescue groups, avoiding newspapers and the like. Be especially careful that reporters don't put words in your mouth.

If a serious full scale rescue is required, a least experienced caver will likely end up being one of the people going out of the cave to get help. After help arrives, he will often be on the surface and can help the total effort by avoiding unnecessary sensational conversation with outsiders and by emphasizing the need to control any crowds of sightseers that may develop. Depending on the type of entrance involved, a large group of people milling around it can be a serious danger both to the injured person and the rescuers, particularly if a pit is involved.

First Aid Kit

A minimum first aid kit for minor injuries should be carried by at least one member of each caving team. It should include several sizes of band aids, a couple of butterfly bandages, disinfectant or first aid cream, and aspirin. Install these in a small plastic or metal box secured with a rubber band or strip of tape. Splints, tourniquets or the more elaborate paraphernalia of advanced first aid are not especially recommended. These are better left to the experts, usually members of an experienced rescue team.

It is also a good idea to have a full first aid kit (of the American Red Cross recommended variety) in the car, along with a blanket and plenty of spare drinking water.

HODAGS

No guide to caving would be complete without mentioning hodags. Now it could be said that cavers are a bit whimsical and that the stories of hodags are no more true than those of the other little people that inhabit the middle regions of the earth. Why, there are even those that say the muddy one-eyed earthworms (also known as cavers) made the whole thing up. But no one who has ever gone caving doubts that hodags are real, at least as real as the one-eyed earthworm himself who also comes to life only in the darkness underground.

For some reason, most hodags, at least the ones that cavers encounter, are of the male variety. At least, they are almost universally named Harry. Anyway, the hairy little creatures seem to spend most of their time making visiting cavers seem unwelcome underground.

They are credited for example, with sneaking up and switching off electric lamps or blowing out carbide lamps just as a caver is stepping over the lip of a 60 foot pit on rappel. In the case of the blown out carbide lamp, it's not too clear if only one hodag is involved, or if it might be, as some cavers report, a whole bevy of the little creatures pushing on a fan-like formation to create one of those strong drafts that unexpectedly sweeps down the passage after hours of calm air (and a prosperous voyage).

It is reliably reported by those cavers who spend a lot of time underground and have managed to befriend a few hodags, that the microfolk really want to be pals with cavers. Except that the evil smells of those flaming eyes and loud clattering noises frighten the sensitive little tykes. But quiet noises, now, that's another matter entirely. Every caver, even those who have never seen a hodag themselves, have heard those *quiet* noises—the whistles, groans, creaks, rumbles, and scrapes heard just down the passage or behind a caver in a crevice. These are hodag sounds, make no mistake about that.

So the next time the caver finds a survey marker moved a few feet from its original position, or feels water rising in a stream passage over his boot tops, or feels a small hand tugging on his ankle as he steps across the bottomless pit (only to find the foot hold used by the others has mysteriously moved), or discovers a well tied knot has become untied a few minutes later, or feels icy water running down his neck when standing in a perfectly dry passage—he'll know that he's not alone after all in this dark miserable hole in the ground—the hodags have come out of their tiny hiding places to keep him company. And have no fear, they'll stay with him for as long as he remains underground, just to be sure the caver gets out, not with dignity, but with dispatch!

11

Sources of Further Information

RECOMMENDED READING

Techniques and Equipment

Butcher, A. L. *Cave Survey.* (Cave Research Group of Great Britain: Publication No. 3). Description of cave surveying methods in Britain.

Cons, David. *Cavecraft.* 184 pp. (London: Harrap, 1966). A primer for the beginning British caver.

Cullingford, Cecil, Ed. *Manual of Caving Techniques.* 416 pp. (London: Routledge and Kegan Paul, 1969). Written by members of Cave Research Group of Great Britain. Very comprehensive and detailed. A veritable text book on techniques and equipment as oriented to the British caver.

Davies, William E. *Cave Maps and Mapping.* (Huntsville, Alabama: Bulletin of the National Speleological Society, September 1947, Number 9, pp 1-7, 37). Somewhat out of date, but still worth reading.

National Speleological Society. *American Caving Accidents 1967; 1968; 1969* and *1970*. Description and analysis of underground accidents, by the NSS Safety and Techniques Committee.

Plummer, William T. *Some Techniques for Cave Exploration*. (Huntsville, Alabama: Bulletin of the National Speleological Society, January 1966, Volume 28, No. 1, pp 22-37). Good survey article, including a bibliography.

Robinson, Donald, *Potholing and Caving*. Paperbound 40 pp. (London: Educational Publications, 1967). One of the Know-the-Game series of beginner books.

Robinson, Donald and Greenback, Anthony. *Caving and Potholing*. 172 pp. (London: Constable, 1964). An excellent introduction for the novice British caver.

Slaven, John. *The Speleoguide*. Paperbound 92 pp. (Visalia, California: Published by the author). A brief but good introduction to caving.

Storey, T. W. *American Caving Illustrated*. 302 pp. (Atlanta, Georgia: Printed by the author). Caving techniques and equipment. Includes several chapters on non-caving subjects like food and cooking, camping and hiking, and first aid.

Thrun, Robert. *Prusiking*. Paperbound. (Huntsville, Alabama: National Speleological Society). A chapter in forthcoming *Caver's Handbook* to be issued in separate sections. Detailed description of knots, mechanical ascenders, and techniques. A must for the vertical caver.

Climbing and Mountaineering

Dixon, C. M. *Rock Climbing*. Paperbound 48 pp. (London: Education Productions, 1958). One of the Know-the-Game series for beginners.

Henderson, Kenneth A. *Handbook of American Mountaineering*. 239 pp. (Cambridge, Massachusetts: Riverside Press, 1942). Now somewhat dated, but still a good basic text on techniques.

Mandolf, Henry I., Ed. *Basic Mountaineering*. Paperbound 112 pp. San Diego, California: Sierra Club, 1961). Good general text. Chapters 1 (Equipment) and 9 (Rock Climbing) are of most value to cavers.

Manning, Harvey. *Mountaineering, The Freedoom of the Hills*. 430 pp. (Seattle: The Mountaineers, 1960). An authoritative book covering all aspects of rock climbing and mountaineering. Best chapters for cavers are Chapter 1 (Equipment) and Chapters 6-11 (Rock Climbing).

Mendenhall, Ruth and John. *Introduction to Rock and Mountain Climbing.* 192 pp. (Harrisburg, Pennsylvania: Stackpole, 1969). Step by step instructions for the beginning rock climber, including snow and ice terrain in addition to rock and mountain techniques.

MIT Outing Club. *Fundamentals of Rock Climbing.* Paperbound 46 pp. (Cambridge, Massachusetts, 1956). Good general advice on belaying, rappelling, knots, and movement.

Sierra Club. *Belaying The Leader—An Omnibus on Climbing Safety.* Paperbound 85 pp. (San Francisco: The Sierra Club, 1956). Good technical information on rock climbing.

U.S. Army. *Mountain Operations.* Paperbound 238 pp. (Washington: Department of the Army, 1959). Most useful chapters for cavers are Chapter 2, pp 125-138 (Rescue Operations) and Chapter 4 (Military Mountaineering).

Wheelock, Walt. *Ropes, Knots, and Slings for Climbers.* Paperbound. (Glendale, California: La Siesta Press, 1967). An excellent survey of the subject.

Cave Exploration and Famous Cavers

Casteret, Norbert. *Ten Years Under the Earth.* (London: The Greystone Press, 1938).

———————— *My Caves.* 172 pp. (London: J. M. Dent and Sons, 1947).

———————— *Cave Men, New and Old.* (London: J. M. Dent and Sons, 1951).

———————— *The Darkness Under the Earth.* 174 pp. (New York: Holt, 1954).

———————— *Descent of Pierre Saint Martin.* (New York: Philosophical Library, 1956).

———————— *More Years Under the Earth.* 164 pp. (London: Neville Spearman, 1962).

Casteret, pioneer cave explorer, recounts many of his underground adventures in this series of books. Fascinating reading.

Chevallier, Pierre. *Subterranean Climbers.* 221 pp. (London: Faber and Faber, 1951). Story of the descent into the Dent de Crolles system in France, among the two or three deepest caves in the world (2,157 feet).

Folsum, Franklin. *Exploring American Caves.* Paperbound, 307 pp. (New York: Collier, 1962). Interesting stories of famous caves, including some information on the formation of caves, cave animals, and techniques. Recommended.

Halliday, William R. *Adventure is Underground.* 206 pp. (New York: Harper, 1959). Story of Western caves and cavers. Recommended.

——————— *Depths of the Earth.* 398 pp. (New York: Harper and Row). More stories of caves and cavers, this time in the entire U.S. Recommended.

Heap, David. *Potholing: Beneath the Northern Pennines.* (London: Routledge and Kegan Paul, 1964). Guide to the pit caves of Britain's Pennine Mountains.

Hovey, Horace Carter. *Celebrated American Caverns.* 228 pp. (New York and London: Johnson Reprint Service, 1970). Originally published in 1896 and long out of print, this fascinating survey of 19th century caves is now available in a beautiful new edition.

Lawrence, Joe Jr. and Brucker, Roger W. *The Caves Beyond.* (New York: Funk and Wagnalls, 1955). The story of the week long 1964 NSS expedition into Floyd Collins Crystal Cave. Out of print, but try your library. Recommended.

Mohr, Charles, E. and Sloane, Howard N., Editors. *Celebrated American Caves.* 339 pp. (New Brunswick, N. J.: Rutgers University Press, 1955). Stories of famous caves and cavers. Recommended.

Stenuit, Robert and Jascinski, Marc. *Caves and the Marvellous World Beneath Us.* 95 pp. (London: Nicolas Vane, 1966). Beautifully illustrated, including color photographs, this book covers cave diving, prehistoric cave art, speleothems, bats, and other cave animals.

Tazieff, Haroun. *Caves of Adventure.* Paperbound 222 pp. (New York: Viking, 1953). Firsthand account of a 2000 foot descent into the Pyrenees.

Conservation

Davidson, J. K. and Bishop, W. A. *Wilderness Resources in Mammoth Cave National Park: A Regional Approach.* (Columbus, Ohio: Cave Research Associates, 1971).

Watson, R. M. and Smith, P. M. *Underness Wilderness: A Point of View.* International Jrl. of Environmental Studies, Vol. 2, pp 217-220, 1971.

National Speleological Society. *Conservation Task Force Report: Guadaloupe; Lost River; Mammoth; Devil's Icebox.* (Huntsville, Alabama 1970/71).

Cave Sciences

Allen, G. A. *Bats.* Paperbound. (New York: Dover, 1962). Excellent treatise on bats.

Barbour, Roger W. and Davis, Wayne H. *Bats of America.* (Lexington, Kentucky: University of Kentucky Press, 1969). An excellent reference book for identifying bats.

Cullingford, C. H. D., Ed. *British Caving—An Introduction to Speleology.* 468 pp. (London: Routledge and Kegan Paul, 1962 — second revised edition). Written by members of the Cave Research Group of Great Britain. Basic authoritative text oriented to British caves and cavers. Some material on equipment and techniques, but companion volume, *Manual of Caving Techniques* by same editor covers these much more thoroughly.

Griffin, D. R. *Listening in the Dark.* (New Haven, Connecticut: Yale University Press, 1958). Acoustics and sonar used by bats to avoid bumping into walls and each other.

Mohr and Poulson. *The Life of the Cave.* (New York: McGraw Hill, 1966). Cave life and ecology. Basic introduction to the subject.

Moore, George W., Ed. *Origin of Limestone Caves.* (Huntsville, Alabama: Bulletin of the National Speleological Society, January 1960, Volume 22, No. 1, pp 1-84). A symposium with discussion.

Moore, George W. and Nicholas, Brother G. *Speleology, the Study of Caves.* Paperbound 120 pp. (Boston: D. C. Heath, 1964). An introductory text, authoritative yet very readable. Highly recommended. Prepared in cooperation with the National Speleological Society.

Siffre, Michel. *Beyond Time.* 228 pp. (London: Chatto and Windus, 1964). Story of 63 days spent underground by a French scientist.

Commercial Caves

Sloane, Howard N. and Gurnee, Russell H. *Visiting American Caves.* 246 pp. (New York: Crown, 1966). A comprehensive guide to all American caves open to the public.

Caves with Prehistoric Art

Grigson, Geoffrey. *The Painted Caves.* 223 pp. (London: Phoenix House, 1957). Description and explanation of the prehistoric art in French and Spanish caves.

Laming, Annette. *Lascaux.* Paperbound 208 pp. (Harmondsworth, England: Penguin, 1959). Description of the cave paintings and the prehistoric people who painted them.

Sieveking, Ann and Gale. *The Caves of France and Northern Spain.* 272 pp. (London: Vista, 1962). A complete and indispensable guide for the visitor to the prehistoric art treasures of France and Spain.

NSS Caving Information Series

1. Information for Neophyte Cavers (2 pp).
2. Techniques of Cave Mapping (2 pp).
3. The Hard Hat (2 pp).
4. Cave Clothing (2 pp).
5. Public Relations and Caving (3 pp).
6. A Micro-Velocity Anemometer (2 pp).
7. Belay Techniques (5 pp).
8. Finding New Caves (2 pp).
9. Belay Tie-Off (2 pp).
10. Tying into a Belay Rope (1 p).
11. Scaling Poles (6 pp).
12. Belaying Rules (1 p).
13. Cable Ladders — Construction, Corrosion, and Safety (5 pp).
14. Building a Wet Suit (5 pp).
15. Bats (2 pp).
16. Standard Map Symbols (2 pp).
17. Cave Rescue (8 pp).
18. Cave Radios in Mapping (7 pp).
19. A Magnetic Induction Cave Radio (5 pp).
20. A Nutritionally Adequate Cavers Ration (9 pp).
21. The Caver's Camera (1 p).
22. Selection of a Caving Camera (7 pp).

(Obtainable from The National Speleological Society, Cave Avenue, Huntsville, Alabama 35810.)

Magazines and Periodicals

Ascent. Sierra Club Mountaineering Journal. Address: 1050 Mills Tower, San Francisco 94104.

NSS Bulletin. A journal of speleology. Detailed in-depth papers and articles on scientific subjects, exploration, and occasionally techniques. Issued quarterly to NSS members.

NSS News. Feature articles, columns, and news about Society events. Issued monthly to NSS members.

Off-Belay. Rock climbing and mountaineering, particularly in the Northwest. Issued every other month. Address: 12416 169th Avenue SE, Renton, Washington 98055.

Summit. Rock climbing and mountaineering. Monthly. Address: P.O. Box 888, Big Bear Lake, California 92315.

Wilderness Camping. For the self-propelled wilderness enthusiast. Six issues per year. Address: 1654 Central Avenue, Albany, New York 12205.

EQUIPMENT SUPPLIERS

The following is mainly a listing of caving and climbing suppliers, but many carry a full line of backpacking, wilderness camping, and related gear. Usually, they will supply a catalog on request. An excellent comprehensive guide covering these sports plus wild water, ski touring, and mountaineering is (price $.50 post paid):

IOCAlums's Products List
Roland and Anne Vineyard
RFD 2, Box 295 (Route 89)
Mansfield Center, Connecticut 06250

ALPINE HUT, 2650 University Village, Seattle, Washington 98105

APEX SAFETY PRODUCTS CO., Washington and Elm Streets, Cleveland, Ohio 44146 (hardhats)

BILL CUDDINGTON, 4729 Lumary Dr., Huntsville, Alabama 35810 (technical caving gear)

BLUE WATER, LTD., P.O. Box 129, Carrollton, Georgia 30117 (caving rope and technical caving gear)

CAMP AND TRAIL OUTFITTERS, 21 Park Place, New York, N. Y. 10007

CAMP TRAILS, 411 W. Clarendon Ave., Phoenix, Arizona 85019

CHOUINARD, Box 150, Ventura, California 93001 (climbing hardware)

DONALD G. DAVIS, P.O. Box 25, Fairplay, Colorado 80440 (Premier carbide lamps)

EASTERN MOUNTAIN SPORTS, 1041 Commonwealth Ave., Boston, Mass. 02215

EDDIE BAUER, P.O. Box 3700, Seattle, Washington 98124 (down bags and jackets)

EDMUND SCIENTIFIC, 101 East Glouchester Pike, Barrington, N. J. 08007 (Nicad batteries)

EIGER, P.O. Box 4037, San Fernando, California 91342 (Edelrid climbing rope)

FIBER-METAL PRODUCTS CO., Chester, Pa. 19190 (hardhats)

FORESTRY SUPPLIERS, INC., Box 8397, Jackson, Mississippi 39204 (Brunton compasses, hard hats, etc.)

FORREST MOUNTAINEERING, Box 7083, Denver, Colorado 80207 (technical climbing gear)

GIBBS PRODUCTS CO., 854 Padley St., Salt Lake City, Utah 84108 (ascenders and technical caving gear)

HOLUBAR MOUNTAINEERING LTD., P.O. Box 7, Boulder, Colorado 80302

JUSTRITE MANUFACTURING CO., 2061 N. Southport Ave., Chicago, Ill. 60614

KOEHLER MANUFACTURING CO., Marlborough, Mass. 01752 (Wheat brand electric lamps)

MINE SAFETY APPLIANCES, 201 North Braddock Ave., Pittsburgh, Pa. 15208 (Electric lamps and hard hats)

MONTICO, ALPINE RECREATION WAREHOUSE, 4-8 Henshaw St., Woburn, Mass. 01801

MOUNTAIN TRAIL SHOP, 4758 Old Wm. Penn Hwy., Murraysville, Pa. 15668

THE NORTH FACE, Box 2399, Station A, Berkeley, California 94702

RECREATIONAL EQUIPMENT, INC., 1525 11th Ave., Seattle, Washington 98122

SIERRA DESIGNS, 4th and Addison St., Berkeley, California 94710

SKI HUT, 1615 University Ave., Berkeley, California 94703

SMILIE CO., 575 Howard St., San Francisco, California 94105

THOMAS BLACK AND SONS, 930 Ford St., Ogdensburg, N. Y. 13669

WELSH MANUFACTURING CO., 4 Magnolia St., Providence, R. I. 02909 (hardhats)

Boot Resoling

MOUNTAIN TRADERS, 1711 Grove St., Berkeley, California 94709

DAVE PAGE, Cobbler, 346 NE 56th St., Seattle, Washington 98105

WHERE TO GO CAVING

The principal caving regions of North America are found in areas where limestone deposits are abundant (for solution caves) or where there has been volcanic activity in geologic times (for lava tubes). Generally speaking this means the Appalachians, southern Georgia and Alabama, northern Florida, the Ozarks, western Texas, eastern New Mexico and northern Mexico, and the lava beds of California and the Northwest. But nearly every state has at least a few caves.

By far the simplest way to find these areas is to look for commercial caves within one's own part of the country. As pointed out, wherever there are commercial caves, there are bound to be some wild caves nearby. (A directory of commercial caves has been listed —see chapter 11, Commercial Caves.)

However, the warning made throughout this book must be repeated, that caving alone is just another form of suicide. Instead of endangering himself, it is recommended that anyone interested in caving but inexperienced find a local chapter of the National Speleological Society or a similar local caving group. Specific information on how to locate such a group is given in Chapter 4. Organized caving groups have the two basic things that every would be caver needs. First, they know how to cave and can teach the novitiate, and second, they know where the caves are.

All of this is another way of saying that a caver is easily damaged by a cave and vice versa, which is why the emphasis is on conservation and safety in this book.

CAVING ASSOCIATION

The National Speleological Society
Cave Avenue
Huntsville, Alabama 35810

This organization is a non-profit society affiliated with the American Association for the Advancement of Science. It has over 100 chapters in all parts of the country. Write to the headquarters at the above address for the location of the closest one. Please include a self-addressed envelope.

Glossary

anchor a secure point (rock, tree, expansion bolt or wedge) to which a caving rope or ladder can be safely attached.
aragonite a less common form of calcium carbonate found in caves (the more common is calcite), usually in the form of needlelike crystals.
bacon or bacon rind speleothem made up of a thin sheet of calcite with alternating bands of color. The brown or darker bands are usually caused by iron oxide.
balanced climbing *see* climbing.
basalt a common type of lava.
biner colloquial for carabiner.
bedding plane the surface or boundary that divides two adjacent beds of sedimentary rock.
belay a method of safe climbing in which a rope is tied around the first caver's chest so that a second caver, who is securely anchored above (or below), can stop a fall. In Britain, known as safetying or life lining.
blowing cave a cave that has large air currents moving in or out for extended periods. Changes in barometric pressure are the cause.
boxwork honeycomb-like speleothem of calcite projecting from a cave wall or ceiling.
bowline a non-slipping loop; the most important single knot any caver must learn.
breakdown large piles of rocks and boulders that have fallen from the ceiling or walls at an earlier time.

cable ladder see ladder.
calcite the most common cave mineral, a crystaline form of calcium carbonate.
carabiner an oval or pear shaped link with a spring-loaded gate in one side, used in climbing and rappelling.
carbide lamp a miner's lamp used by cavers. Acetylene gas produced by mixing water with carbide is ignited and burns at a jet positioned in a reflector.
carbonic acid a weak acid made from carbon dioxide and rain or soil water that slowly dissolves limestone to form caves.
cave a natural void beneath the earth, usually made up of several rooms and passages, large enough to enter.
cave formation *see* speleothem.
cave ice year round ice formed in some lava caves.
cave spring *see* resurgence.
chimney 1) a vertical or near vertical shaft, either tubular or simply where two walls come close together. 2) climbing up or down such a shaft by means of pressure on both surfaces by back and feet.
climbing caving movement used in pits, fissures, and cave walls; involves three points of contact and other classic rock climbing techniques.

column a speleothem formed where a hanging stalactite and a rising stalagmite have grown together.

corkscrew passage a twisting passage, often a tight crawlway, either horizontal or vertical.

crevice a narrow opening in the floor of a cave, often 10 to 40 feet or more deep; also a high narrow passage.

dead cave a cave in which the speleothems have stopped growing because water is no longer seeping into the cave.

dig an attempt to break into a cave or a new area of a known cave by excavation, and in some cases blasting.

doline *see* sink.

dolomite a sedimentary rock similar to limestone, in which caves can occur.

domepit a large dome-shaped cavity above a room or passage, caused by solution, not breakdown.

drapery a thin curtain-shaped speleothem caused by a sheet of dripping water rather than a single series of drops.

dripstone any of several calcite deposits caused by dripping water, including stalactites, stalagmites, and flowstone.

duckunder a place in a stream passage or lake room where the ceiling comes down into the water causing a caver to duckunder for a few feet. A longer duckunder is called a siphon. Both are extremely hazardous and definitely not for beginners.

expansion bolt a type of anchor for ropes and ladders. It is placed into a hole drilled into limestone and expands as it is driven in. Typically, a ¼ inch bolt will support about 2,000 lbs.

exposure a dangerous condition caused by extreme wet and cold where body heat is being lost rapidly. Lack of food can accelerate the effects. Exposure can kill if not checked. Also called hypothermia.

false floor a thin floor made of calcite or lava under which dirt or gravel has been worked away.

fill clay, mud, rock, or other material found on the floor of a cave.

fissure a narrow crack, break, or fracture. Fissures (and crevices) are usually negotiated by chimneying or traversing movements.

flowstone a coating of calcite deposited by flowing water.

flute, stream scallop-like ripples in a cave wall caused by stream action.

formation *see* speleothem.

Gibbs ascender a device used for prusiking up a rope.

grotto a small room or chamber opening off of a larger one. Also a local chapter of the National Speleological Society.

guano in cave terminology, bat dung. A very rich fertilizer.

gypsum a sedimentary rock (primarily calcium sulphate), which is softer and more soluble than limestone. Caves can occur in gypsum.

gypsum flowers and hair varieties of delicate gypsum speleothems, often of great beauty.

hand line a short (10 to 20 foot) fixed rope used for climbing or scrambling on steep pitches when holds are scarce.

helectite a beautiful twisting speleothem that seems to grow in defiance of the laws of gravity.

hold, hand or foot small ledge, knob, or crevice that can support the hand or foot as an assist in climbing, scrambling, or chimneying.

hydrology scientific study of underground and surface water.

hypothermia *see* exposure.

ice cave a type of cave, usually in lava, which contains ice all year.

joint a crack, usually formed at right angles to the bedding plane in limestone.

Jumar ascender a device for prusiking up a rope.

junction a place where two or more passages come together.

karst terrain with many sinkholes, disappearing streams, underground drainage, and caves.

karabiner *see* carabiner

keyhole a keyhole shaped passage, usually tight.

ladder caving ladder, made of aircraft cable and aluminum rungs, used for climbing out of 10 to 30 foot pits. Deeper pits are now more commonly prusiked.

lava tube a type of cave formed in lava as it cools. Often nearly circular in cross section.

lead a side passage, hopefully leading to more cave.

live cave a cave with speleothems still being developed by water.

limestone a rock composed primarily of calcium carbonate and readily dissolved by carbonic acid. Most caves are formed in limestone.

lost river a stream that runs underground for some of its length.

marble limestone later subjected to heat and pressure. Many caves occur in marble.

moon milk putty-like form of flowstone.

nut, climbing *see* wedge.

pit a roughly tubular hole, also loosely applied to crevices and fissures. A number of caves have pit entrances resulting from breakdown, dome pit solution, or stream action.

pitch a steep ascent or descent.

pinch-out in caver's talk, a passage that tapers down and becomes too small to penetrate.

ponor the point where a stream disappears in karst country. Also called a swallet or swallow hole.

prusik knot a basic caving knot used primarily for climbing up a fixed rope. Easily moved upward, but when weight or tension is applied, it holds fast.

prusiking method of climbing out of a pit, crevice, or fissure using prusik knots or mechanical ascenders on a fixed rope with slings attached to feet and chest. Used for pits from 25 to 1,000 feet or more.

rappelling method of safely sliding down a fixed rope using carabiners with brake bars, a rappel rack, or spool. Today, cavers usually rappel into a pit, then come out via ladder or prusik.

resurgence point where a cave stream reappears on the surface.

rimstone a crusty calcite deposit at the edge of a lake or small series of pools.

safety *see* belay.

scramble a half crawling/half climbing movement used to negotiate steep or muddy slopes.

sea cave a void or cavity in rocks along a shore caused by wave action.

shelter cave a cavity in any kind of rock offering shelter from the weather.

sink a depression, often 30 to 100 feet or more across, caused by collapse of a cave passage below the surface.

sink hole a hole in the bottom or side of a sink that leads into a cave.

siphon a long duckunder requiring underwater swimming, and often some type of breathing device like Scuba gear. Very hazardous. Not for beginners.

snaplink *see* carabiner.

sling short (eight to 20 foot) piece of flat nylon material (or 5/16 or 3/8 inch diameter climbing rope), used to attach ropes and ladders to anchors, and safety lines, rappelling devices, and ascenders to cavers.

soda straw a thin, hollow form of stalactite, through which drops of water descend and deposit calcite at the bottom.

solution tube type of passage developed by dissolving action. Usually of nearly tubular proportions.

squeeze an extremely tight passage. Also squeezeway.

speleogenesis the origin and development of caves.

speleothem generic name for all cave deposits of calcite, aragonite, and gypsum. Includes stalactites, stalagmites, columns, drapery, flowstone, rimstone, gypsum flowers, helectites, and others caused by deposition.
stalactite a hanging speleothem formed by calcite in dripping water.
stalagmite a calcite deposit built upward from the floor by dripping water.
swallet or swallow hole *see* ponor.
talus cave an underground cavity formed by falling rock.
troglodyte an animal, including the human, who lives in a cave.
travertine a course type of flowstone, formed by calcite in water flowing over a surface.
tube *see* solution tube and lava tube.
walk-in cave a cave with a large entrance, often, but not always, with large walking passages inside.
water table top or highest level of underground water in a given area. Below this level, all cavities and voids are flooded.
wedge, climbing a new type of easy-to-place and easy-to-remove anchor for ropes and ladders. It is inserted into a crack or small crevice and acts like a chock stone. Most have a hole in the center to attach a rope or sling. Also called a climbing nut.
wild cave an undeveloped cave in its natural state, in contrast to a commercial cave where lighting and paths have been added.